AF682393

Disaster Recovery

Disaster Recovery

Editor

Sidharth Saini

scitus
academics

Disaster Recovery

Edited by **Sidharth Saini**

Printed in 2017

ISBN: 978-1-68117-129-6

Library of Congress Control Number: 2015936550

© 2016 by

SCITUS Academics LLC,
616, Corporate Way, Suite 2, 4766,
Valley Cottage, NY 10989

www.scitusacademics.com

This book contains information obtained from highly regarded resources. Copyright for individual articles remains with the authors as indicated. All chapters are distributed under the terms of the Creative Commons Attribution License, which permits unrestricted use, distribution, and reproduction in any medium, provided the original author and source are credited.

Notice

Reasonable efforts have been made to publish reliable data and views articulated in the chapters are those of the individual contributors, and not necessarily those of the editors or publishers. Editors or publishers are not responsible for the accuracy of the information in the published chapters or consequences of their use. The publisher believes no responsibility for any damage or grievance to the persons or property arising out of the use of any materials, instructions, methods or thoughts in the book. The editors and the publisher have attempted to trace the copyright holders of all material reproduced in this publication and apologize to copyright holders if permission has not been obtained. If any copyright holder has not been acknowledged, please write to us so we may rectify.

Contents

Preface

Disaster recovery (DR) involves a set of policies and procedures to enable the recovery or continuation of vital technology infrastructure and systems following a natural or human-induced disaster. Disaster recovery focuses on the IT or technology systems supporting critical business functions, as opposed to business continuity, which involves keeping all essential aspects of a business functioning despite significant disruptive events. Disaster recovery is therefore a subset of business continuity.

Editor

Chapter 1

Short- and Long-term Evacuation of People and Livestock during a Volcanic Crisis: Lessons from the 1991 Eruption of Volcán Hudson, Chile

Thomas Wilson[1], Jim Cole[1], David Johnston[2], Shane Cronin[3], Carol Stewart[4], and Andre Dantas[1]

[1]Natural Hazards Research Centre, Geological Sciences, University of Canterbury, Christchurch, New Zealand

[2]Joint Centre for Disaster Management, GNS/Massey University, Lower Hutt, New Zealand

[3]Volcanic Risk Solutions, Massey University, Palmerston North, New Zealand

[4]Private consultant, 37 Harland Street, Brooklyn, Wellington, New Zealand

ABSTRACT

Human and livestock evacuation during volcanic crises is an essential component of volcanic risk management. This study investigates the evacuation of human and livestock populations from areas impacted by ashfall from the 1991 Hudson eruption, Patagonia. The eruption was one of the largest in the 20th century resulting in significant impacts on rural communities in affected areas, including the evacuation of people and livestock. In the short-term (<3 months), evacuation of people from farms and rural towns was driven primarily by ashfall and ash storm impacts on public health and essential services. Severe impacts on livestock and the inability to restore vegetation growth following pasture burial, also meant pastoral farming became unsustainable in the short term. This resulted in evacuation of farms for usually <1, but up to 4 years following the ashfall and subsequent intense ash-storms. In areas of very heavy ashfall (>1 m) or where agricultural systems were stressed (from drought and long-term low commodity prices) many farms were abandoned, resulting in permanent migration of the farm population. Farms and farmers under pressure from marginal economic returns were the least likely to cope with the 'shock' of the ashfall. The financial capacity of farmers was important in their resilience and ability to return once conditions improved, although emotional attachment to the land sometime outweighed financial considerations.

Evacuation of livestock in areas affected by ash falls was undertaken by many farmers, but it was not very successful or economically justifiable. Access for livestock trucks to the impacted area was difficult due to a poor road network, ashfall and snow induced blockage, and remobilised ash inhibiting visibility. The lack of reliable records of livestock populations inhibited evacuation and efforts to supply supplementary feed to the remaining livestock. The very poor condition of livestock prior to the eruption and burial of feed following the eruption often made evacuation uneconomic as well as reducing livestock resilience to cope with the eruption and transport impacts. The lack of capacity within the local livestock market and lack of available grazing land for the influx of transported livestock were also key failings of the evacuation effort.

INTRODUCTION

Volcanic eruptions produce a range of hazards that can endanger life and have wider societal impacts such as on agricultural activities, requiring mass evacuation of human and livestock populations. Mass evacuation from a particular area is necessary when volcanic hazards threaten the safety of those within the area (pre-impact) or following the impact of a hazard which has subsequently rendered the area uninhabitable (post-impact). Volcanic eruptions may also continue for years to decades causing on-going damage and risk to life which curtail recovery efforts and require the long-term evacuation of communities. Many human evacuations during volcanic crises have been recorded globally (see Table 1 for examples). Witham (2005) estimated over 5 million people were evacuated or affected by 248 volcanic events during the 20th century. Advances in volcanic risk management, particularly volcano surveillance (monitoring), which can provide timely evacuation warnings have dramatically reduced casualties from volcanic hazards over the past 20 years (Witham 2005).

Table 1: Selected human and livestock evacuations during volcanic crises

Eruption	Area Impacted	Details	Reference
1886 Tarawera eruption	Bay of Plenty, New Zealand	An estimated 20,000 livestock were evacuated from areas receiving >10 mm of ashfall.	Keam 1988
1973 Eldfell eruption	Heimaey, Iceland	Within 6 hours of the eruption nearly all of Heimaey's 5,300 residents had been evacuated to the Icelandic mainland. Several hundred sheep were evacuated to the mainland, whilst remaining cattle and poultry were slaughtered	Williams & Moore 1983;Wilson 2009

1976-77 Soufrière eruption	Guadeloupe, West Indies	Approximately 73,500 people were temporarily evacuated, mostly from Plymouth, the island's capital.	McGuire et al. 2009
1980 Mt St Helens eruption	Washington, United States	35 property owners at Spirit Lake evacuated and an exclusion zone established.	The Daily News/ The Journal America 1980
1986 Nevado del Ruiz eruption	Tomila, Columbia	15,000 people evacuated in low-lying areas due to lahar hazards (in the aftermath of the destruction of Armero town by lahars following a small eruption in 1985).	Voight 1989
1990-1995 Unzen eruption	Shimabara, Japan	At the peak of the crisis, over 160,000 people used short-term evacuation accommodation, and nearly 5,669 used temporary housing for a period of up to four and half years due to pyroclastic flow hazards.	Shimizu et al. 2007
1991 Pinatubo eruption	Luzon, Philippines	Over 200,000 people were evacuated from around Pinatubo due to pyroclastic flow, heavy ash fall and lahar hazards. Significant subsequent lahar hazards resulted in on-going evacuations by returnees.	Newhall et al. 1997

1995-present Soufriere Hills eruption	Montserrat, West Indies	Approximately two thirds of the ~12,000 island's inhabitants were evacuated between 1995-97. On-going activity has led to their long-term displacement.	Clay et al. 1999;McGuire et al. 2009
1999 Tungurahua eruption	Tungurahua, Ecuador	Enforced evacuation of the entire population (16,000) from Baños due to potential pyroclastic flow, lahar and ashfall hazards.	Tobin & Whiteford 2002;Lane et al. 2003
2002 Nyiragongo eruption	Goma, Democratic Republic of Congo	Lava flows forced the evacuation of ~300,000 people from Goma city and surrounds and left 120,000 homeless.	UNDP 2004
2006 Merapi eruption	Yogyakarta, Indonesia	Over 20,000 people evacuated from western and southern flanks of Merapi due to pyroclastic flow and lahar hazards. Livestock remained on farms within the evacuation zone and tended to by farmers travelling in and out during the day. Many farmers attempted to sell livestock following feed destruction and isolation from their farms.	Wilson et al. 2007

2008-present Chaiten eruption	Northern Patagonia, Chile	Preventative evacuation of ~5,000 people occurred from proximal areas due to potential pyroclastic flow and lahar hazards, particular Chaiten town. Some evacuation occurred in rural areas from heavy ashfall hazards. Over 20,000 cattle were evacuated from areas impacted by heavy ashfalls in the weeks following the eruption.	Lara 2009

Wilson *et al. Journal of Applied Volcanology* 2012 1:2 doi: 10.1186/2191-5040-1-2

Mass evacuations during volcanic crises are commonly used to mitigate proximal volcanic hazards, such as pyroclastic flows, lahars, volcanic gases, landslides and lava flows (Table 1). However, volcanic ashfall may also impact human and animal health, water supplies, critical infrastructure, buildings, and agricultural activities which in turn drives evacuations (Cook et al. 1981; Blong 1984;Baxter 1986; Cronin et al. 1998; Neild et al. 1998; Johnston et al. 2000; Stewart et al. 2006; IVHHN 2007; Wilson & Cole 2007). Additionally, remobilisation of pyroclastic deposits by water and wind can cause regional scale impacts to exposed populations motivating prolonged evacuation and ultimately relocation of communities from an area. An obvious example was the evacuation and ultimately relocation of thousands of people in central Luzon, Philippines, due to on-going, widespread lahars following the 1991 Pinatubo eruption (Janda et al. 1996). The extreme impacts that a volcanic eruption may cause, may be the catalysts for social change depending on the pre-existing economic, social and political environments, accelerating the rate at which adjustments in social and political institutions occur (Blong 1984; Tobin & Whiteford 2002). However, beyond the Pinatubo eruption, there are few studies examining post-impact evacuations driven by acute primary ashfall and chronic secondary reworking of pyroclastic deposits; and how this has affected the long-term recovery of a region.

Similarly, there have been few accounts of livestock evacuation during volcanic crises. This is despite agriculture being a common land-use in volcanic areas (Table 1) and livestock having significant economic and emotional value to farmers (Annen & Wagner 2003; Jenner 2007; Wilson et al. 2009). There is scarce information examining whether the evacuation of livestock increases the motivation of farmers to evacuate (Wilson et al. 2009). Despite other studies suggesting that the willingness or reluctance to evacuate are determined by factors which include: an individual's perception whether evacuation will lead to a positive outcome; fears that property (e.g. livestock and agricultural produce) will be destroyed or stolen if left behind; and attachment to place (e.g. Blong 1984; Mileti et al. 1991; Lindell and Perry 1992; Cola 1996; Dibben & Chester 1999; Tobin & Whiteford 2002;LEGS 2009). With increased livestock trading and transport in modern agricultural systems, the mass evacuation of livestock is considered a popular option by farmers (Wilson et al. 2009). It is a controversial management response following recent volcanic crises, where it was rejected as a possibility by agricultural officials during the 1995/96 Ruapehu eruptions in New Zealand (Neild et al. 1998; Wilson et al. 2009), but was utilised within days of the 2008 eruption of Chaiten, Chile (T. M. Wilson, *unpublished field data*).

There is growing international recognition that during disaster response saving livelihoods through support to key livelihood assets, in addition to saving people's lives, can have a significant bearing on short- and long-term recovery outcomes for livestock dependent or reliant communities. A key example is the Livestock Emergency Guidelines and Standards (LEGS) project which aims to increase the quality of emergency response when livestock are key social and economic assets for the affected community by promoting international guidelines and minimum standards for the design, implementation and assessment of livestock interventions during disasters (LEGS 2009). Society has also increased its awareness of zoological vulnerability for livestock and is significantly less tolerant of widespread livestock losses in the aftermath of a disaster. Beyond the obvious economic loss, the significant psychosocial impacts to individuals when valued animals are lost or harmed, and the negative reputational impacts to region's and countries following mass fatalities of livestock has demanded increasing attention from emergency managers (Irvine 2009).

This study investigates the evacuation, relocation and return of human and livestock populations in areas impacted by ashfall from the 1991 Hudson eruption, Patagonia. It seeks to interpret the drivers causing farmers to evacuate and in some cases abandon farms, the agronomic and social impacts of the evacuation, the rationale and execution of livestock evacuation, and the factors that encouraged farmers to return to evacuated or abandoned farms. Field work was completed between 20 January and 8 February 2008 in the region of southern Patagonia impacted by the 12-15 August 1991 ashfall (Figure 1). Farming communities were visited in a transect from the upper Río Ibáñez valley in Chile, which received >1 m of ashfall, to the distal portions of the ash plume at the Atlantic coast of Santa Cruz province, Argentina (Figure 1). Field methods included semi-structured interviews with 32 farmers, and 11 municipal officials or agricultural experts (e.g., veterinarians or agricultural field officers) who had experienced the 12-15 August 1991 eruption or participated in the response and recovery operations.

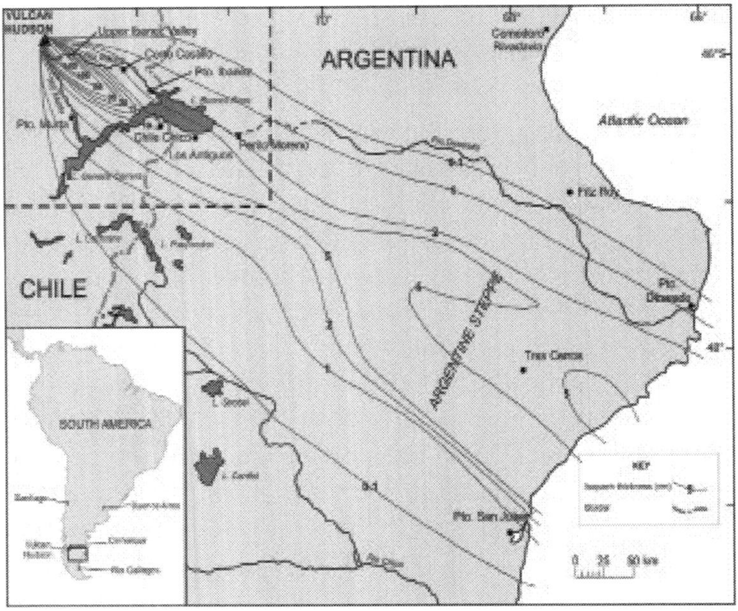

Figure 1: Location map and isopach map of eruptive deposits of the 12-15 August 1991 Hudson eruption (after Scasso et al. 1994). Figure 2 is shown as a dashed box.

THE 1991 ERUPTION OF VOLCÁNO HUDSON

Volcán Hudson (45°54′ S; 72°58′ W) is part of the Chilean Southern Volcanic Zone (33-46°S) (Kratzman et al. 2008). At least 12 Holocene explosive eruptions have occurred at Hudson, the most significant of which were eruptions 6,700 years before present (yrs BP), 3,600 yrs BP, and in 1991 (Naranjo & Stern 1998). The 1991 eruption consisted of two separate, partially sub-glacial phreatoplinian explosive phases on 8-9 August and 12-15 August 1991. It was the second phase of the eruption, on 12-15 August, 1991, that dispersed ash on a narrow, elongated ESE sector of Patagonia, covering a land area of more than 100 000 km^2 (Figure 1) and caused most of the evacuations.

The eruption produced 4.3 km^3 bulk volume (2.7 km^3 dense rock equivalent) of tephra fall deposits, making it one of the largest explosive eruptions of the 20th century (Kratzmann et al. 2008). The elongated shape of the deposit was the result of strong northwest winds.

WHERE DID EVACUATION OCCUR?

The area impacted by ashfall was climatically and ecologically diverse, resulting in a wide range of agricultural practices. The Ibáñez valley, in the southern Andean mountain range, west of Puerto Ibáñez supported low to medium intensity pastoral farming of sheep and some cattle (>700 mm of annual rainfall). The area surrounding Lago Gen. Carerra/Buenos Aires (Figure 2) has irrigated valleys supporting intensive pastoral and horticultural farming at Puerto Ibáñez, Chile Chico, Los Antiguos and Perito Moreno (200-500 mm annual rainfall). The vast arid steppe region that extends east to the Atlantic coast (Figure 1) supports low intensity extensive pastoral farming of sheep (<200 mm of annual rainfall) (Peri & Bloomberg 2002). The ashfall buried short winter pastures, forcing pastoral farmers to use what limited supplementary feed was available. Heavy snowfall also coincided with the ash deposition in the Andean regions. Farms were already overstocked, due to poor commodity prices for meat and wool in the previous season, adding further stress. Hence, in general, livestock were already in poor condition. Following ashfall, many suffered from starvation, and other health problems due to ingestion of ash with feed as well as being directly impacted by ashfall, which resulted in widespread livestock deaths (Rubin et al. 1994; Wilson 2009). Valdivia (1993) estimated >1 million sheep and several thousand cattle died in Chile and Argentina following the ashfall. High winds common to the area remobilised ash deposits, creating billowing clouds of fine ash (referred to as ash storms), which abraded and repeatedly covered pasture re-growth, exacerbating the impact (Wilson et al. 2011).

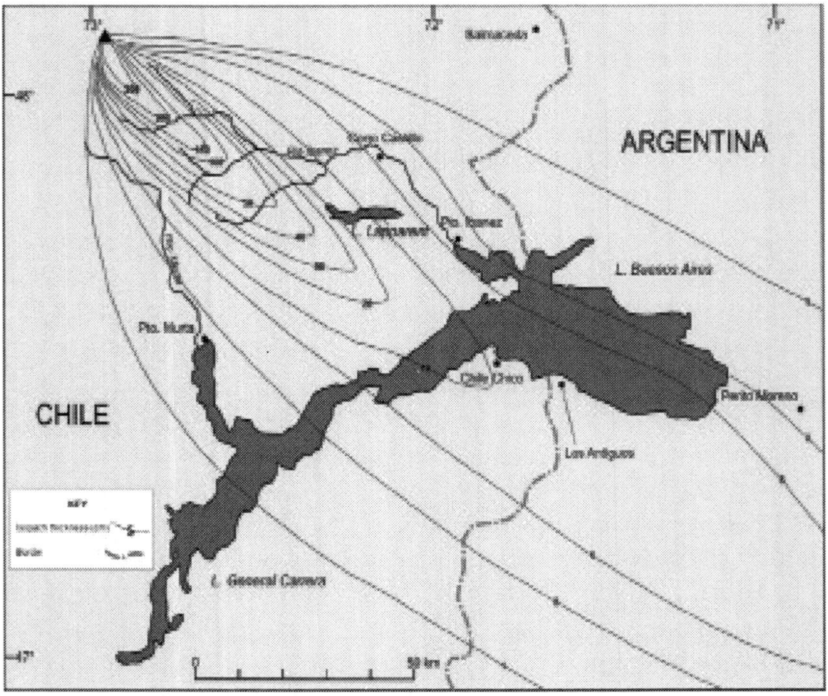

Figure 2: Location map and isopach map of the Ibanez River catchment (after Scasso et al. 1994).

Horticultural farms (in the irrigated valleys) were not initially affected, as crops and fruit trees were in winter dormancy or had yet to be planted. However subsequent ash storms during the crucial spring growth period (September to November) caused significant damage and loss of yields. People on farms and the small rural service towns in the impacted area had significant health problems due to the airborne ash during ashfalls and particularly during subsequent ash storms.

Farmers initially evacuated to the closest towns, such as Puerto Ibáñez, Chile Chico, Los Antiguos, Perito Moreno, Puerto San Julian and Tres Cerros, but eventually moved further away in search of employment. Many headed to regional centres such as Coihaique and Commodoro Rivadavia (Figure1). Farmer evacuation from the three main agricultural areas impacted by ashfall is investigated below.

Ibáñez Valley

The initial focus of emergency managers responding to rural communities exposed to heavy ashfall was human safety. The Ibáñez valley received very heavy ashfalls (500->2000 mm; Figure 1;Naranjo et al. 1993) and most farmers evacuated, with many self-evacuating and others using army and ONEMI (Chilean Civil Defence) vehicles. The farming families generally left within hours or days of the eruption onset. Farms were small (100-200 ha) and carried low numbers of livestock, typically cattle with stocking rates of ~0.3/ha. The factors that led to interviewed farmers deciding to evacuate included: homestead roof collapse, or fear it may collapse (Table 2); concerns about health impacts from the ash, including homestead contamination and severe ongoing problems with wind-blown fine ash; fears of further ashfalls; lack of livestock feed following pasture burial and exhaustion of supplementary feed. Farmers described heavy ashfall and snow accumulations on homestead roofs requiring frequent cleaning to prevent collapse, particularly in areas that received >600 mm of ash (Table 2), which agrees with previous known instances of roof collapse from ashfall (Spence et al. 2005). Several farmers reported large lapilli breaking glass windows. In one case this occurred ~50 kilometres from the volcano. Homesteads were rapidly contaminated with ash. The limited available feed was also rapidly covered in thick (>200 mm) ash. Veterinarian Don Julio Cerda Cordero of Servicio Agrícola y Ganadero (SAG; a branch of the Chilean Ministry of Agriculture) reported that in extreme cases, ash ingestion with feed caused gastro-intestinal blockages. Starvation and gastro-intestinal complications from ash consumption were also reported as the primary cause of livestock mortality (Rubin et al. 1994). There were few reports of contaminated water or water shortages as a significant problem for either people or animals, mostly because surface water supplies were spring fed and cleared quickly.

Table 2: Roof collapse mitigation actions

Ash & snow fall depth (farmer estimate)	Ashfall depth (Naranjo et al. 1993)	Distance from vent	Details
150 cm	90 cm	36 km	Roof collapsed on wooden constructed homestead, despite initial cleaning attempts
100-120 cm	35 cm	42 km	Regular cleaning required to prevent collapse on wooden construction homestead
80-100 cm	40 cm	73 km	Wooden constructed homestead & farm building roofs collapsed under weight of ash and snow - not cleared due to evacuation
50-60 cm	15 cm	54 km	No structural damage to wooden homestead. Regular roof cleaning due to fears of collapse.

Wilson *et al. Journal of Applied Volcanology* 2012 1:2 doi: 10.1186/2191-5040-1-2

Most farmers interviewed had evacuated from the Ibáñez valley by 15 August. The remaining evacuees left by 30 August. Some farmers remained in government welfare centres and emergency accommodation in towns such as Cerro Castillo for up to 4 years, or relocated out of the area in search of work. Most farmers returned regularly to assess the vegetation recovery. Many farmers said they were very reluctant to leave their farms, citing strong financial and emotional attachment to their land and livestock.

Irrigated Valleys

High intensity mixed horticultural and pastoral farms (stocking rate >3/ha) and horticultural farms (typically fruit orchards) were located in river valleys with developed irrigation networks. They typically experienced between 50-120 mm of ashfall at Puerto Ibáñez, Chile Chico, Los Antiguos and Perito Moreno. These farms used cultivation and fertiliser improvement, permanent willow and poplar tree wind breaks, and were located close to population centres. Whilst not catastrophically impacted by ashfall, many mixed and horticultural farms were evacuated for weeks to several years, in response to health hazards (especially from wind remobilised ash) and uncertainty around future eruption impacts (Wilson 2009). One interviewee had evacuated from Los Antiguos due to the health hazards from airborne ash inflaming a chronic asthma condition. The family had evacuated for several months, during which most of their 70 sheep died due to starvation after ash buried pasture. Some farmers wished to evacuate due to health hazards, but remained on their farms because of fears of their houses being looted, the need to care for their livestock, and had missed any aid subsequently offered. Short-term evacuations of towns occurred during ash storms 1-6 months following the ashfall. Over 600 women and children evacuated from Chile Chico, Los Antiguos and Perito Moreno (Wilson et al. 2011).

Residents in Puerto Ibáñez, Chile Chico and Los Antiguos all reported cleaning ash from their roofs, fearful that they may collapse. Over 100 adobe-style buildings in Chile Chico suffered structural damage from the weight of ash on their roofs. Some totally collapsed, whilst many others had large cracks in load-bearing walls. Light snow and rainfall greatly increased the weight of the ash exacerbating the impacts. Most interviewees said they used shovels to remove ash from their roofs, because it was like cement and very heavy when it became wet.

Puerto Ibáñez municipal officials reported that there has been a significant population decline in the town and surrounding area, down from 4,000 in 1991 to 2,700 in 2001; a level at which it has since stabilised. In contrast to the upper Ibáñez valley area, widespread farm abandonment did not occur at Puerto Ibáñez with ash depths of 50-80 mm being initially more manageable. However, on-going ash-storms repeatedly damaged pastures, crops and livestock, and reduced soil fertility, causing agricultural recovery to be extremely slow and challenging (Wilson et al. 2011).

Steppe

Farms in the Argentine steppe are typically very large (often tens of thousands of hectares), run mostly sheep, and at very low stocking rates

(<0.1/ha). This region experienced up to 80 mm of ashfall, but more typically fall deposits were between 1-50 mm (Scasso et al. 1994). This was enough to cover sparse feed resources and fill shallow watering holes. Extreme effects from wind remobilised ash exacerbated and prolonged these problems (Wilson et al. 2011). The fine ash (<0.5 mm) from primary ashfall and from ash storms penetrated houses, making them "uninhabitable". Farmers in the region evacuated into towns, such as Chile Chico, Los Antiguos, Perito Moreno and Puerto San Julian due to household contamination and health fears. Impacts on livestock were extreme, with starvation, acute dehydration, and tooth abrasion leading to widespread and severe livestock losses (Wilson et al. 2011; Rubin et al. 1994). Livestock watering holes were contaminated with primary and windblown ash, creating muddy "death traps" for starved and exhausted livestock. Livestock were dispersed across the large farms, making them difficult to muster for supplementary feeding and provision of fresh water. Ash-storms reduced visibility and made movement difficult, compounding recovery problems. The large steppe farms had few technological improvements, such as irrigation or wind breaks to mitigate impacts or assist recovery (Wilson 2009). The poor financial state of farmers limited any capacity to absorb losses and forced abandonment of many farms in the ensuing months, particularly as ash-storms continued to bury or destroy scarce feed.

A substantial human evacuation out of remote, extensive farming areas around Chile Chico, Los Antiguos, Tres Cerros and Puerto San Julian was reported following the ashfall and subsequent ash-storms (because of farm abandonments there were few farmers available to interview In the Argentine coastal steppe - therefore information was obtained by interviewing municipal officials and the former Mayor of Puerto San Julian). Farmers evacuated "temporarily" due to remobilised ash from strong winds and fine ash contaminating house interiors. However, this ultimately became permanent abandonment, because ash storms continued to damage vegetation and livestock. Farmers were forced to lay-off workers and ultimately abandon their farms. As many as 80% of extensive pastoral farmers in the steppe are thought to have abandoned their farms due to productivity losses and health effects from wind-blown ash, according to former Perito Moreno agricultural officer Shaquib Hamer. Farmer and mechanic Don Hugo Ciselli estimated 70-80% of people had left the Tres Cerros area (~100 people). He said approximately 50 farms have been abandoned in the area, but not sold, although recently (2008) 2-3 farmers have returned. Former Mayor of Puerto San Julian, Alberto James Alder reported that approximately 7,000 people were forced from their farms in the local area. Farmers evacuated immediately after and headed for town or relatives. Most went back periodically to check on their farms but ultimately realised sheep farming was no longer viable. These farms had been struggling prior to the eruption with dry conditions and low wool and meat commodity prices. The eruption and on-going ash-storms appeared to have accelerated their decline.

LIVESTOCK EVACUATION

A significant animal welfare crisis rapidly developed during and following the ashfall, with farms overstocked and the scarce feed buried under ash and snow. Farmers rapidly exhausted limited supplementary feed stocks, whilst Chilean government agricultural agencies SAG and the Instituto de Desarrollo Agropecuario (INDAP) attempted to organise and assist with evacuating livestock and sourcing further supplementary feed. Livestock evacuation operations began in the Ibáñez valley area almost immediately and later in Chile Chico. Livestock trucks were used to bring in supplementary feed into the affected area including over 100,000 kg of molasses and 30,000 bales (30 × 40 × 120 cm) of hay. An estimated 5,000 cows and 3,000 sheep were evacuated out from all affected areas (Mario Christian Moreno, Regional Secretary, INDAP, pers. comm. 2008).

Interviewed SAG and INDAP officers recalled that there was significant confusion during the initial stages of the agricultural emergency response, particularly with respect to what should be done and how severe the impact was. Lack of an accurate livestock census was a significant deficiency for emergency response planning. The number of livestock thought to be in the area was significantly higher than official figures, which led to a major underestimation of the number of livestock requiring feeding and evacuation. Sourcing additional supplementary feed from surrounding regions was extremely difficult, due to the time of year and recent dry conditions. Logistical coordination problems impeded supplementary feed reaching staging posts in the disaster zone. Transporting the feed to impacted areas was also difficult, as ashfall covered roads and mixed with snow, making traction very difficult. Wind-blown ash inhibited visibility. Trucks were only operated on roads with thin to moderate ash falls (<50 mm of compacted ash), but the main access road into the Ibáñez valley became covered in heavy ash deposits making traction difficult. Rain and snow in mid-August turned the ash to thick mud, making access impossible for heavy vehicles, and required livestock to be walked out to staging points (Don Julio Cerda Cordero, *pers. comm.* 2008). Falling and remobilised ash dramatically cut visibility, making it difficult to drive. Evacuation/feeding operations had to be halted for several days from 15 August due to the poor conditions. Sheep and cows abandoned on farms chased vehicles, hoping they would be fed, and sometimes blocked road traffic. Ash deposits had compacted sufficiently by mid-September so that trucks could access farther up the Ibáñez valley (Don Julio Cerda Cordero *pers. comm.* 2008). However, the main evacuation efforts only lasted until early September (Mario Christian Moreno, Regional Secretary, INDAP, pers. comm. 2008). The Chilean government assisted by paying for supplementary feed, but did not assist with transportation costs (Don Julio Cerda Cordero, *pers. comm.* 2008).

The decision to evacuate livestock was in the hands of individual farmers and dependent on their financial means. Most were unable to evacuate. Some farmers even halted and unloaded their livestock from trucks, after government agricultural agencies had loaded their livestock onto trucks. There was no threshold of ash depth that dictated the necessity for evacuation. In some cases, light ashfalls contaminated water supplies forcing evacuation. In other cases there may have been significant thicknesses of ashfall, but with access to sufficient clean water and supplementary feed, animals could survive (Don Julio Cerda Cordero *pers. comm.* 2008). Cows were generally evacuated preferentially due to their higher value. Farmers were required to find a location where the livestock could go, so only those with contacts outside the impact zone were able to evacuate their livestock. When combined with the poor access to impacted areas and high animal mortality rates, only a low percentage of livestock in the heavily impacted areas were moved. Farmers who did evacuate their livestock were often later forced to sell them at very low prices to avoid paying for hired grazing. Often this did not cover the evacuation transport costs.

Uncertainty of health impacts on livestock, livestock feed availability, medium term pasture recovery and the threat of future eruptions motivated livestock (particularly cattle) evacuation from Chile Chico by ferry to Puerto Ibáñez and later by truck across the border at Los Antiguos, driving around Lago Buenos Aires and back into Chile. Argentine border controls were relaxed to assist the evacuation of people and livestock. Blocked roads meant livestock could not initially be evacuated from Los Antiguos, leading to severe losses in sheep herds from starvation once supplementary feed was exhausted. When roads reopened livestock feed was brought in and some animals transported for sale.

The speed of evacuation was crucial for animal welfare. Livestock were already in poor condition, and began dying within days of feed being covered by ash. Some evacuated livestock died during or immediately following evacuation due to gastro-intestinal complications, usually starvation related.

IMPACT OF FARM ABANDONMENT

Most farmers initially believed they would only evacuate their farms for a short-period (days to weeks). In the weeks, months and years after the eruption it became evident that wind-blown ash and household contamination (leading to health concerns) were continuing hazards. In many areas pastoral farming was no longer viable until vegetation recovered. Large scale livestock deaths and evacuations led to significant de-stocking of farms and otherwise compromising productivity. Land values were significantly depreciated, livestock assets all but lost, and with continuing impacts from ash storms, farmers were commonly forced from their lands, particularly in the Ibáñez valley and in the steppe region. Farmers leaving the area attempted to sell their farms, but struggled to find buyers or refused to accept the very low prices offered, so abandoned their farms but planned to return later. The Chilean government purchased some badly impacted farms in the worst affected areas of the Ibáñez valley, but there was no such support for farmers in the Argentinean steppe (Wilson 2009). This created a significant social problem, because the impacts drastically reduced farmer's financial equity and since most only knew sheep farming, they had few other employment prospects. All interviewed farmers said they had always expected to return to their farms, citing strong family ties and emotional attachment. In general, elderly farmers were more likely to evacuate their farms and were less likely to return, citing they did not have the physical or emotional energy to recover from the eruption event. The impact of the loss of financial equity was exacerbated and mentally crushing for displaced elderly farmers, who either expected to retire on their land or use the money from the sale of their farm to support retirement.

The death of animal populations on a large scale was also very difficult for pastoral farmers to cope with (Hall et al. 2004; Jenner 2007). Social and economic distribution from evacuation and reaching the difficult decision of whether to abandon farms compounded stress, and supports initiatives such as LEGS (2009).

Why did they Return?

Many abandoned farms had been re-occupied in the Ibáñez valley by 2008. However, few had been returned to in the steppe region, despite having experienced thinner ashfalls. A key factor for returning was the sufficient recovery of vegetation. In the Ibáñez valley with increasing ash thickness, the duration of abandonment increased (Figure 3). Within 1-5 years many farms close to Cerro Castillo were inhabited again (50-70 mm ashfall). By 1994-95 reworking of ash by wind and water had in places slowed and had also exposed the original soil surface (Wilson 2009). In the upper Ibáñez valley with a total coverage of >500 mm ash, farms were abandoned for up to 10 years. Eventual vegetation recovery was aided by wind removing ash overburdens, and of spreading of seed and hay on the ash deposit to re-establish pastures. Despite re-habitation by 2008, agricultural production was extremely low compared to pre-eruption condition. Areas with greater than 1.5 m ashfall did not support any farming in 2008. The poor chemical and physical fertility characteristics of the ash inhibit pasture re-establishment. Stabilisation of ash deposits against wind erosion was an important factor to allow vegetation recovery, especially in the dry, windy steppe region (Wilson 2009).

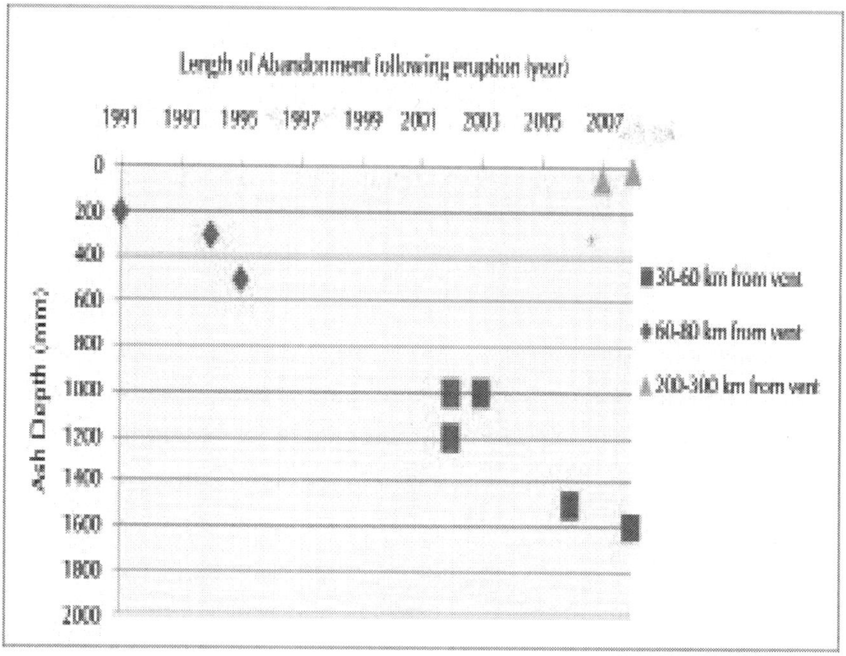

Figure 3: Time till farm reoccupation following the 1991 Hudson eruption (after Wilson 2009).

Factors influencing farmers' return included financial capacity, a decrease in health fears with the reduction of ash storms, and the strength of emotional attachment to the land. Farmers found it a challenge to raise the necessary funds to bring farms back into production, despite credit assistance available from government agricultural agencies, such as INDAP in Chile. In the steppe region, the huge cost of re-stocking and re-developing farms was a significant barrier. The three farmers interviewed who maintained farms either at Tres Cerros or Puerto San Julian maintained significant other business interests to provide financial security. Some farms diversified and changed land use practices. In the upper Ibáñez valley with ash >1 m, farmers mostly relied on lumber and fire wood extraction and carried few livestock. Farms were planted in exotic pine forestry in some areas with 1200 mm of ash. There was no evidence of land disputes between farmers or resettlement unrest occurring at Hudson, contrasting with the aftermath of the Parícutin eruption, where murders and suicides occurred (Rees 1979).

LIMITATIONS OF THE STUDY

This is one of the first studies to study the fate of human and livestock populations following long term evacuation due to a large explosive volcanic eruption. It has acknowledged limitations; for example much of the information contained within this paper is of a qualitative nature, derived from semi-structured interviews. This approach allowed detailed and in-depth data collection on complex issues. But there are several limitations of this approach which must be acknowledged, including repeatability, potential for interviewer bias, relevance of qualitative data collected, difficulties in generalising one-on-one interviews, and validity (e.g. if the respondent is biased). In addition, there are few interviews with people who have permanently moved away from the impacted area. It is therefore not possible from this study to analyse the motivation to leave or the relative success, whether it be economic, social, psychological or otherwise, of those who have returned and those who have permanently left the region. Such data would be invaluable for shaping future disaster management planning.

This study highlights the importance of environmental (e.g. drought and windy conditions) and human systems context (e.g. low commodity prices) preceding and following the disaster event. However, one of the greatest limitations of this study is the lack of longitudinal demographic and socioeconomic data to more robustly analyse spatial and temporal changes to population and economic productivity following the 1991 ashfall. Wider sociological aspects, such as incidence of domestic violence and mental health problems over time, would also be of interest. Whilst we did not record any accounts of evacuated people causing antagonism when they moved to new communities outside the impact zone, as might have been anticipated (due to competition for resources for example; Lindell and Perry 1992; Hewitt 1997). Few interviews were in areas which had received evacuated people and sufficient time may have passed for this to have been forgotten.

CONCLUSIONS

The impacts from the ashfall and ash storms resulted in evacuations out of the area in the short and long term. The following types of responses were identified from interviews:

- Short-term (<3 months) evacuation of farms and rural towns due to ashfall and ash storm impacts on public health and essential services;
- Evacuation of farms for usually <1, but up to 4, years following ashfall and subsequent intense ash-storms, which reduced farm productivity in the short term by burying feed;
- Abandonment of farms and permanent migration (usually following farm sale) from areas that suffered a long-term decline of agricultural income due to destroyed vegetation and soil from ashfall and ash-storm effects, drought and falling commodity prices (usually in the steppe);

Widespread farm abandonments (>3 years) occurred in two areas with different climates, agricultural practices and which received very different ashfall depths; the Ibáñez valley (500->2000 mm), and the steppe region of extensive pastoral ranch style farming (<75 mm). Household contamination and health concerns from ashfall and ash storms motivated human evacuations from farms and rural towns up to hundreds of kilometres from the erupting volcano. Destruction of farm resources, such as pastures, soils and animals and inability to restore feed production, meant pastoral farming became unsustainable. This often turned the evacuation into abandonment, with farming only viable once vegetation recovered. The financial capacity of farmers was important in their resilience and in enabling their return once conditions improved (LEGS 2009). Farms and farmers under pressure from marginal economic returns were less likely to cope with the 'shock' of the ashfall. On-going impacts from ash storms and the financial requirements necessary to restock farms and invest in soils and pastures created further barriers to return to farming. Some farmers maintained other financial interests during this transition.

It is important to recognise the substantial loss of equity farmers suffered in their land and livestock value within days of heavy ashfall. This created major social problems and led to significant migration from the region. The most vulnerable were elderly farmers relying on the equity to assist retirement. Farmers who returned cited strong family ties, emotional attachment to the area, and residual economic value in the land as factors driving their decision, despite the hardships and low economic returns.

It has been recognised that people may not evacuate if there is no clear planning for their pets or livestock (Zeigler et al. 1981; Heath et al. 2001a;b; Hall et al. 2004; Hunt et al. 2008; LEGS 2009). Concerns for livestock dictated some farmer's actions when considering evacuation, but ultimately the death of animals, fear of roof collapse, and health concerns motivated most to evacuate. The evacuation of livestock was a secondary priority to human evacuation, but since livestock represent significant economic value to impacted farmers, there may be strong desire and even economic justification to remove livestock from an impact area. High livestock populations of large animals such as cattle create significant logistical problems, especially if road access is limited (Wilson et al. 2009). Evacuation of livestock around Hudson was not very successful nor economically justifiable. Livestock truck access to the impacted area was difficult due to limited roads, ashfall and snow induced blockage, and remobilised ash inhibiting visibility. The lack of reliable records of Chilean livestock populations severely inhibited evacuation and efforts to supply supplementary feed. The poor condition of livestock often made evacuation uneconomic as well as reducing the resilience of livestock to cope with the eruption impacts and the stresses of transport. The lack of capacity within the local livestock market and lack of available grazing for the influx of transported livestock were important failings of the evacuation effort.

AUTHORS' CONTRIBUTIONS

Please use the following format as an example, including using initials to refer to each author's contribution: TW carried out the research design, undertook the fieldwork and draft the manuscript. JC assisted in research design, carried out fieldwork and assisted in drafting the manuscript. DJ conceived of the study and participated in its design and coordination. SC assisted in the research design and significant revision of the manuscript. CS assisted in research design and carried out fieldwork. AD assisted in draft the manuscript. All authors read and approved the final manuscript.

ACKNOWLEDGEMENTS

This research was supported by the New Zealand Ministry of Agriculture & Forestry Operational Research Grant SUS7802 (TW), the Foundation for Research Science and Technology grants MAUX0401 (SC) and C05X0804 (JC & DJ) and the Earthquake Commission.

REFERENCES

1. Annen C, Wagner J (2003) The Impact of Volcanic Eruptions During the 1990s. Natural Hazards Review 4(4):169-175
2. Baxter PJ (1986) Preventative health measures in volcanic eruptions. American Journal of Public Health 76:84-90
3. Blong RJ (1984) Volcanic hazards: a sourcebook on the effects of eruptions. Academic Press, Australia. p 424
4. Clay E, Barrow C, Benson C, Dempster J, Kokelaar P, Pillai N, Seaman J (1999) An evaluation of HMG's response to the Montserrat volcanic emergency. UK Department of International Development Evaluation Report EV635. London Two volumes. 86-182
5. Cola RM (1996) Respnoses of Pampanga Households to Lahar Warnings: Lessons from two villages in the Pasig-Patrero River Watershed. In: Newhall CG, Punongbayan RS (eds) Fire and Mud, Philippine Institute of Volcanology and Seismology and University of Washington Press, Quezon City and Seattle.
6. Cook RJ, Barron JC, Papendick RI, Williams GJ (1981) Impact of Agriculture of the Mount St. Helens Eruptions. Science 211:16-22
7. Cronin SJ, Hedley MJ, Neall VE, Smith RG (1998) Agronomic impact of tephra fallout from the 1995 and 1996 Ruapehu Volcano eruptions, New Zealand. Environmental Geology 34(1):21-30
8. Dibben C, Chester DK (1999) Human Vulnerability in Volcanic Environments: the Case of Furnas, São Miguel, Azores. Journal of Volcanology and Geothermal Research 92(1):133-150
9. Hall MJ, Ng A, Ursano RJ, Holloway H, Fullerton C, Casper J (2004) Psychological Impact of the Animal-Human Bond in

Disaster Preparedness and Response. Journal of Psychiatric Practice 10(6):368-374

10. Heath SE, Beck AM, Kass PH, Glickman LT (2001a) Risk factors for pet evacuation failure after a slow-onset disaster. Journal of the American Veterinary Medical Association 218(12):1905-1910
11. Heath SE, Voeks SK, Glickman LT (2001b) Epidemiologic features of pet evacuation failure in a rapid-onset disaster. Journal of the American Veterinary Medical Association 218(12):1898-1904
12. Hunt M, Al-Awadi H, Johnson M (2008) Psychological sequelae of pet loss following Hurricane Katrina. ANTHROZOOS 21(2):109-121
13. IVHHN (2007) The Health Hazards of Volcanic Ash - A guide for the public. International Volcanic Health Hazard Network pamphlet. 16
14. Janda RJ, Daag AS, de los Reyes PJ, Newhall C (1996) Assessment and response to lahar hazard around Mount Pinatubo, 1991 to 1993, in Newhall. In: CG, Punongbayan RS (eds) Fire and mud: Eruptions and lahars of Mount Pinatubo, Philippines: Quezon City, Philippine Institute of Volcanology and Seismology, and Seattle, University of Washington Press. pp 107-139
15. Jenner M (2007) The psychological impact of responding to agricultural emergencies. The Australian Journal of Emergency Management 22(2):25-31
16. Johnston DM, Houghton BF, Neall VE, Ronan KR, Paton D (2000) Impacts of the 1945 and 1995-1996 Ruapehu eruptions, New Zealand: An example of increasing societal vulnerability. Geological Society of America Bulletin 112(5):720-726
17. Keam RF (1988) Tarawera: the volcanic eruption of 10 June 1886 A.D., RF Keam Physics Department, University of Auckland, New Zealand.
18. Kratzmann DJ, Carey S, Scasso R, Naranjo J (2008) Compositional variations and magma mixing in the 1991 eruptions of Hudson volcano, Chile. Bulletin of Volcanology 71(4):419-439
19. Lane LR, Tobin GA, Whiteford LM (2003) Volcanic hazard or economic destitution: hard choices in Banos, Ecuador. Environmental Hazards 5:23-34

20. Lara LE (2009) The 2008 eruption of the Chaitén Volcano, Chile: a preliminary report. Andean Geology 36(1):125-129
21. LEGS (2009) Livestock Emergency Guidelines and Standards. Practical Action Publishing, Warwickshire. p 252 Available from: <http://www.livestock-emergency.net >
22. Lindell MK, Perry RW (1992) Behavioral Foundations of Community Emergency Planning. Hemisphere Press. Washington DC.
23. McGuire WJ, Solana MC, Kilburn CRJ, Sanderson D (2009) Improving communication during volcanic crises on small, vulnerable islands. Journal of Volcanology and Geothermal Research 183:63-75
24. Mileti DS, Bolton PA, Fernandez G, Updike RG (1991) The Eruption of Nevado Del Ruiz Volcano Colombia, South America, November 13, 1985. Commission on Engineering and Technical Systems (National Academy Press). Washington, D.C.
25. Naranjo JA, Stern CR (1998) Holocene explosive activity of Hudson Volcano, southern Andes. Bulletin of Volcanology 59:291-306
26. Naranjo JA, Moreno H, Banks N (1993) La erupcion del volcan Hudson en 1991 (46ºS), region XI, Aisen, Chile. Sernageomin, Santiago, Chile 44:50
27. Neild J, O'Flaherty P, Hedley P, Underwood R, Johnston D, Christenson B, Brown P (1998) Impact of a Volcanic Eruption on Agriculture and Forestry in New Zealand. MAF Policy Technical Paper 99/2. 101
28. Newhall C, Hendley JW, Stauffer PH (1997) The Cataclysmic 1991 Eruption of Mount Pinatubo, Philippines, U.S. Geological Survey Fact Sheet. 113-97
29. Peri PL, Bloomberg M (2002) Windbreaks in southern Patagonia, Argentina: A review of research on growth models, windspeed reduction, and effects on crops. Agroforestry Systems 56:129-144
30. Rees JD (1979) Effects of the eruption of Parícutin Volcano on landforms, vegetation, and human occupancy. In: Sheets PD Grayson DK (ed) Volcanic activity and human ecology, New York, Academic Press. pp 249-292

31. Rubin CH, Noji FK, Seligman PJ, Holtz JL, Grande J, Vittani F (1994) Evaluating a fluorosis hazard after a volcanic eruption. Archives of Environmental Health 49(5):395-397

32. Scasso RA, Corbella H, Tiberi P (1994) Sedimentological analysis of the tephra from the 12-15 August 1991 eruption of Hudson volcano. Bulletin of Volcanology 56:121-132

33. Shimizu H, Matsushima T, Nakada S, Fujii T, Takarada S, Miyabuchi Y, Ui T, Miyake Y, Sugimoto T, Nagai D (2007) Intra-meeting excursion - Unzen eruption - Disaster and recovery - Fieldtrip guide, Cities and volcanoes 5 conference. Shimabara.

34. Spence RJS, Kelman I, Baxter PJ, Zuccaro G, Petrazzuoli S (2005) Residential Building and Occupant Vulnerability to Tephra Fall. Natural Hazards and Earth System Sciences 5:477-494

35. Stewart C, Leonard GS, Horwell CJ, Thordarson T, Cronin SJ (2006) Contamination of water supplies by volcanic ashfall: A literature review and simple impact modelling. Journal of Volcanology and Geothermal Research 158:296-306

36. The Daily News/The Journal America (1980) Volcano - The Eruption of Mount St. Helens. Longview Publishing Company, Longview. p 96

37. Tobin GA, Whiteford LM (2002) Community Resilience and Volcano Hazard: The Eruption of Tungurahua and Evacuation of the Faldas in Ecuador. Disasters 26(1):28-48

38. UNDP (2004) Volcano Risk Reduction: A Case Study from Goma (DRC). Disaster Reduction Unit, United Nations Development Programme - Bureau for Crisis Prevention and Recovery.

39. Valdivia VHVDI (1993) Analisis Geografico del Volcan Hudson Y Sus Tipos de Riesgo. In: Geografia y (ed) Tesis presentada para optar al titulo de Profesor de Historia, Civica. Universidad Austral de Chile.

40. Voight B (1989) The 1985 Nevado del Ruiz volcano catastrophe: anatomy and retrospection. Journal of Volcanology and Geothermal Research 42:151-188

41. Williams RS, Moore JG (1983) Man against volcano: The eruption on Heimaey, Vestmannaeyjar, Iceland: U.S. Geological Survey General Interest Publication. 27

42. Wilson TM, Cole JW (2007) Potential impact of ash eruptions on dairy farms from a study of the effects on a farm in eastern Bay of Plenty, New Zealand; implications for hazard mitigation. Natural Hazards 43(1):103-128

43. Wilson T, Kaye G, Stewart C, Cole J (2007) Impacts of the 2006 eruption of Merapi volcano, Indonesia, on agriculture and infrastructure. GNS Science Report.2007/07 69

44. Wilson TM, Dantas A, Cole JW (2009) Modelling livestock evacuation following a volcanic eruption in New Zealand: An example from Taranaki volcano. New Zealand Journal of Agricultural Research 52:99-110

45. Wilson TM (2009) Vulnerability of Pastoral Farming Systems. Unpublished Doctorate of Philosophy Thesis. In: University of Canterbury Library. Christchurch, New Zealand.

46. Wilson TM, Cole JW, Stewart C, Cronin SJ, Johnston DM (2011) Ash Storm: Impacts of wind remobilised volcanic ash on rural communities and agriculture following the 1991 Hudson eruption, southern Patagonia, Chile. Bulletin of Volcanology 73:223-239

47. Witham CS (2005) Volcanic disasters and incidents: A new database. Journal of Volcanology and Geothermal Research 148:191-233

48. Zeigler DJ, Brunn SD, Johnson JH (1981) Evacuation from a Nuclear Technological Disaster. Geographical Review 71(1):1-16

Chapter 2

Characteristics of Meteorological Disasters and Their Impacts on the Agricultural Ecosystems in the Northwest of China: A Case Study in Xinjiang

Meihua Wu[1, 2], Yaning Chen[2], Huaijun Wang[2,] and Guili Sun[3]

[1]Key Laboratory of Oasis Ecology, School of Resource and Environmental Sciences, Xinjiang University, Urumqi 830046, Xinjiang, China

[2]State Key Laboratory of Desert and Oasis Ecology, Xinjiang Institute of Ecology and Geography, Chinese Academy of Sciences (CAS), CAS, No. 818 South Beijing Road, Urumqi 830011, Xinjiang, China

[3]Forestry and Horticulture Department, Xinjiang Agriculture University, Urumqi 830052, Xinjiang, China

ABSTRACT

Background

In recent years, the meteorological extreme events have caused the direct economic losses and human mortality increased significantly. While there has been a paucity of information regarding trends in meteorological disasters in Xinjiang. Based on two extreme climate measurements, i.e., the Palmer Drought Severity Index (PDSI) and the agricultural disaster area, the influence of meteorological disasters on agriculture were analyzed during the period 1960-2010.

Results

Temperature extremes exhibited patterns consistent with warming, with a large proportion of stations having statistically significant trends. The warming trends in the indices derived using daily minimum temperatures were greater than those obtained using maximum temperatures. Most of the precipitation indices exhibited increasing trends across the region, and the increased precipitation was due to the increase in both precipitation frequency and intensity. (2) The drought indices increased significantly in most regions of Xinjiang, and the seasonal PDSI exhibited significant correlations with the annual PDSI. For the entire geographical study area, two contrasting periods were evident in the PDSI between 1961 and 2010. Wet conditions dominated from 1987 to 2010, whereas persistent drought conditions occurred from 1960 to 1986. (3) Increased climate extremes resulted in increased agricultural disaster area. During warm summers, the droughts intensified; the corresponding snowmelt flood also became stronger. In addition, the sharply reduced effective irrigation area exacerbated the increased agricultural disaster area.

Conclusions

Climate change has affected the local agricultural oasis ecosystem and the yield and quality of crops in Xinjiang, leading to increased instability in agricultural production.

BACKGROUND

According to the World Meteorological Organization (WMO) estimate, losses caused by meteorological disasters have accounted for 85% of the total losses cause by natural disasters (Qiang et al. [2001]; Qing [2008]). In the last 20 years, the direct economic losses caused by meteorological extreme events have increased exponentially, and human mortality has also significantly increased (Loukas et al. [2010]). China is one of the countries that has been most severely affected by the meteorological disasters around the world due to its complex terrain conditions, leading to higher frequencies of extreme weather compared to other countries (Zhang et al. [1991]). The crop disaster area caused by various meteorological disasters has been as high as 5×10^7 hectares every year, and the population affected by major meteorological disasters, such as typhoons, rainstorms, droughts, heat waves, and sand storms, has reached 4×10^8 (Qing[2008]). During the period 1990–2006, the direct economic losses caused by meteorological disasters in China was 1859×10^8 RMB per year, accounting for an average of for 2.8% of the annual GDP (Qing [2008]).

Xinjiang has complex natural conditions that lead to a high frequency of various types of strong natural disasters (Ye and Chen [1996]). Meteorological disasters and their derivative disasters have accounted for 83% of the total loss due to all natural disasters, and the number of deaths has accounted for 85% of the total (Liu [1995]). These disasters have also exacerbated the deterioration of the ecological environment. Some new characteristics of these meteorological disasters and their derivative disasters have been found since the 1980s, including the increase in the types of events, increased frequency, and increased strength. According to incomplete statistics, the average annual loss was 1×10^7-5×10^7 RMB, 1.4×10^8 RMB, and 20×10^8 -50×10^8RMB during the period 1950–1970, during the 1980s, and since the 1990s, respectively, accounting for $2 \sim 3.5\%$ of the GDP in Xinjiang. Since the 1980s, the drought disaster area, inundated area, and various economic losses have increased significantly (Xu et al. [2008]; Bai et al. [2012]; Chen and Gao [2010]; Chen et al. [2008]).

Xinjiang is located in the arid region of northwestern China; its ecological environment is very fragile. The recovery period of the agricultural ecological system to disasters is relatively long. Analyzing the effects of meteorological disasters on the agricultural ecosystem in Xinjiang is very important for formulating corresponding disaster prevention countermeasures that can protect the ecological and environmental security. In addition, such an analysis is also favorable for formulating agricultural adaptation and mitigation measures in response to climate change and for maintaining social stability, which are very important for promoting the sustainable development of the national economy.

Description of study area

Xinjiang Province (Figure 1) is located in northwestern China. With an area of 16.61×10^8 km^2, Xianjiang is the largest province in China. Xinjiang is a typical inland arid region. The Altai Mountains, Tian Mountains, and Kunlun Mountains extend from north to south in Xinjiang. The Junggar Basin in North Xinjiang and the Tarim Basin in South Xinjiang are located between these three mountains. Xinjiang is divided into southern and northern areas by its natural landscape, resulting in two mountain-basin systems with different hydrological and thermal conditions (Xu et al. [2004]). The unique mountain-basin landscape is responsible for the frequent occurrence of natural disasters. Alpine areas are distributed widely in Xinjiang and possess numerous glaciers and permanent snow, which increase the frequency of outburst floods. The middle area experiences the highest precipitation in Xinjiang (Gao et al. [2002]), and water erosion is very strong; therefore, floods and snowfalls result in most of the extreme events (Hong and Adler [2008]). In the low mountain areas, active fault zones increase the occurrence of floods, landslides, and debris flows during rainstorms and seasonal snowmelt (Sun et al. [2014]).

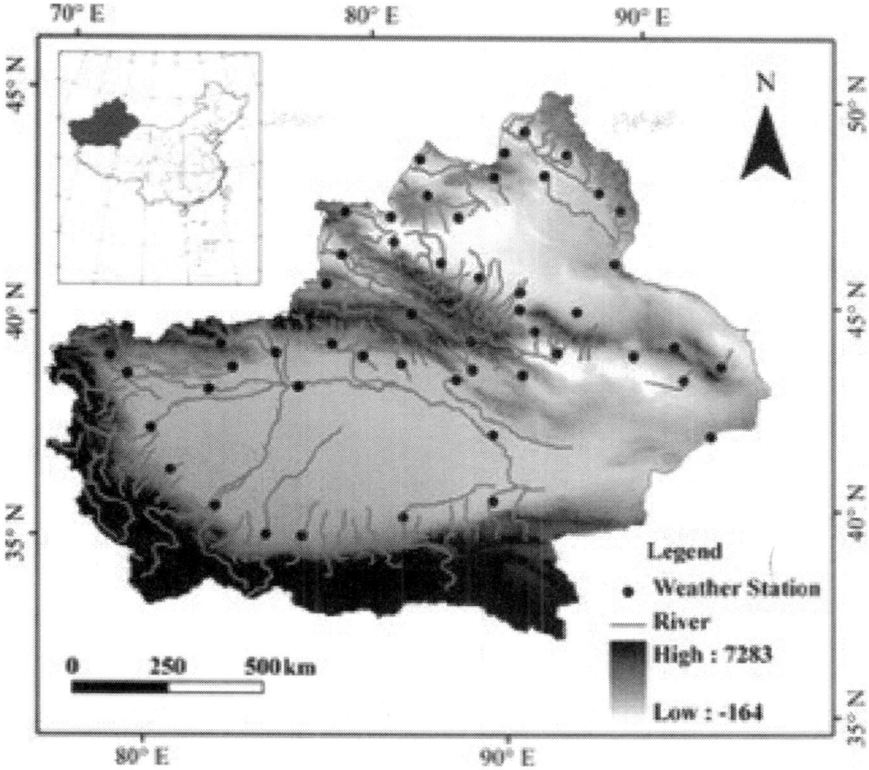

Figure 1: Study area and weather stations in Xinjiang Province.

Methods

Daily data, include the maximum temperature, minimum temperature, precipitation, relative humidity, sunshine hours, air pressure, and wind speed (1 January 1960 to 28 February 2011) covering the Xinjiang region were provided by the National Climate Center (NCC) of the China Meteorological Administration (CMA). For this region, 53 stations passed the internal homogeneity check of the China National Meteorological Center (CNMC), including a moving t test (Peterson et al. [1998]), standard normal homogeneity test (Alexandersson [1986]), and departure accumulating method (Buishand [1982]). Data were analyzed using the RclimDex package (software and documentation available for download from http://etccdi.pacificclimate.org). Table 1 lists the indices used in this study. The regional average is the arithmetic mean of all stations in Xinjiang. The Xinjiang disaster area data were derived from the China Statistical Yearbook and Chinese agricultural statistics. The effective irrigation area and rural power data were also collected from the China Statistical Yearbook and Chinese agricultural statistics.

Table 1: Definitions of 6 temperature indices and 5 precipitation indices used in this study; all of the indices are calculated using RClimDex

ID	Indicator name	Definitions	UNITS
Cold extremes			
TN10p	Cool nights	Percentage of days when TN < 10th percentile	Days
TNn	Coldest nights	Monthly minimum value of daily minimum temp	°C
CSDI	Cold spell duration indicator	Annual count of days with at least 6 consecutive days when TN < 10th percentile	Days
Warm extremes			
TX90p	Warm days	Percentage of days when TX > 90th percentile	Days
TXx	Warmest days	Monthly maximum value of daily maximum temp	°C
WSDI	Warm spell duration indicator	Annual count of days with at least 6 consecutive days when TX > 90th percentile	Days
Variability extremes			
DTR	Diurnal temperature range	Monthly mean difference between TX and TN	°C

Wu *et al.*

Wu *et al. Geoenvironmental Disasters* 2015 **2**:3 dloi: 10.1186/s40677-015-0015-8

The PDSI is widely used in drought evaluation studies, which was developed by Palmer ([1965]) to measure the cumulative departure in atmospheric moisture supply and demand. The PDSI not only accounts for precipitation but also accounts for temperature, which has a large effect on evapotranspiration and soil moisture (Liu et al. [2012]). The

PDSI soil parameter that is used for bucket water balance is the available water content (AWC). The AWC was determined from the State Soil Geographic Database (STATSGO) for the top 100 cm of the soil profile. However, a common critique of the PDSI is that the behavior of the index at various locations is inconsistent, making spatial comparisons of the PDSI difficult. The SC-PDSI automatically calibrates the behavior at any location by replacing empirical constants with dynamically calculated values. More details regarding the PDSI and SC-PDSI can be found in the works of Liu et al. ([2012]) and Wells et al. ([2004]).

We used a nonparametric Kendall's tau-based slope estimator in this study (Sen 1968); statistical significance for the trends in extreme climate indices was determined using the Mann-Kendall test (Kendall [1975]; Mann [1945]). A trend was considered to be statistically significant if it was significant at the 5% level. The results of the M-K test are substantially affected by serial correlation; therefore, we adopted the Yue and Pilon method, which uses the R package "ZYP" to remove lag-1 autocorrelations (Yue et al. [2002]).

RESULTS AND DISCUSSION

Characteristics of Climate Extremes

Figure 2 depicts the spatial trends in the temperature indices for cold extreme (Figure 2a-b). For cold nights (TN10), 51 stations (96%) had significant decreasing trends, with a regional trend of −1.89 days/decade. The temperature on the coldest nights (TNn) also exhibited a statistically significant increase at approximately 70% of the stations, and the regional trend was 0.7°C/decade. The cold spell duration indicator (CSDI) (not shown) decreased at a rate of −0.7 days/decade, while only 25% of the studied stations exhibited a significant change.

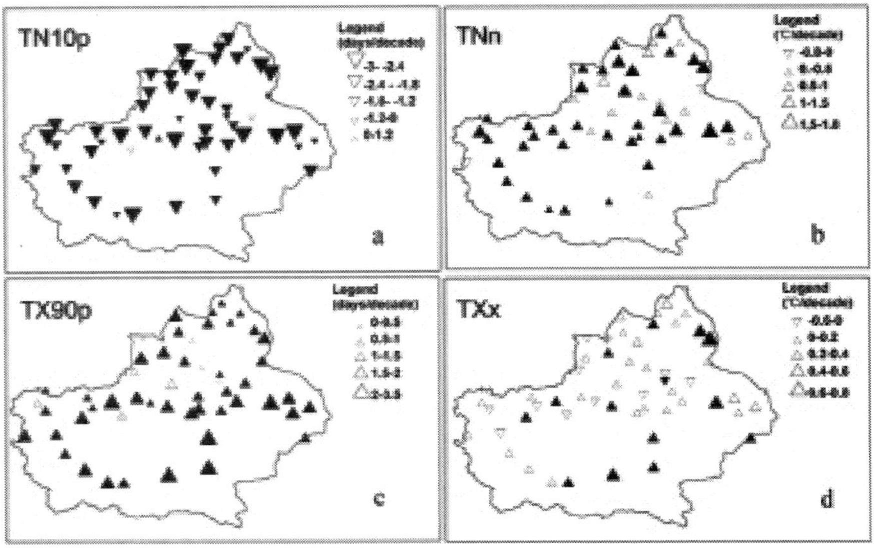

Figure 2: Spatial patterns of the changes in temperatures extremes per decade during the period 1960–2010 in Xinjiang.a: annual TN10p; b: annual TNn; c: annual TX90p; d: annual TXx. Positive/negative trends are shown as up/down triangles; the filled triangles are related to statistically significant trends (significant at the 0.05 level). The triangle sizes are proportional to the magnitude of the trends.

For the warm extremes (Figure 2c-d), 83% of the stations exhibited statistically significant trends for warm nights (TX90); the regional trend for TN90 was 1.25 days/decade. The warmest days (TXx) also exhibited an increasing trend, while only 25% of the stations exhibited significant trends. The warm spell duration indicator (WSDI) (not shown) also increased, and the regional trend was 1.98 days/decade.

The regional seasonal trends in the DTR are shown in Figure 3. The regional trend in the annual DTR was −0.28°C/decade. Winter exhibit the most significant decreasing trend, with a value of −4°C/decade (Figure 3d). The decrease in the DTR was because the minimum temperature increased faster than the maximum temperature.

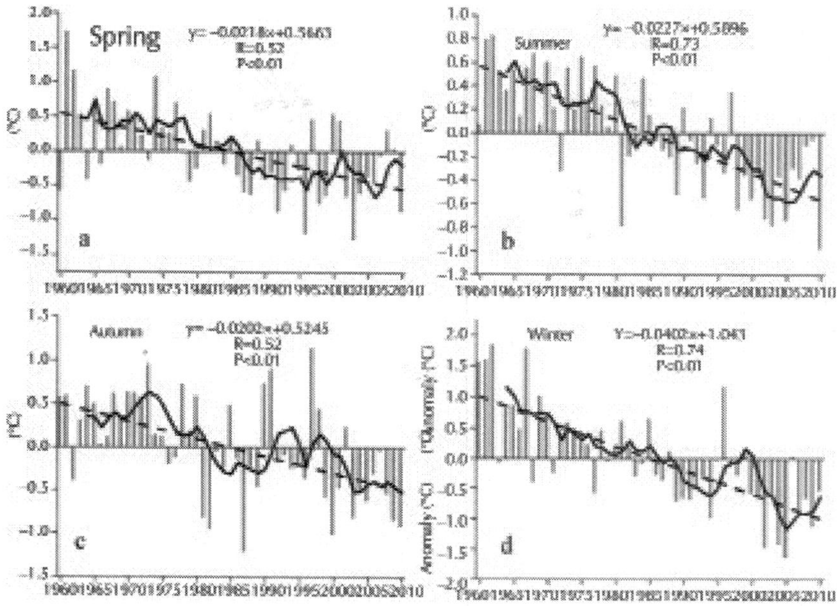

Figure 3: Regional trends in the DTR (the column is the annual anomaly series; the solid line is the 5-year-smoothed average; and the dashed line is the linear regression).a: spring DTR; b: summer DTR; c: autumn DTR; d: winter DTR.

The spatial distribution of the trends in precipitation extremes is show in Figure 4. For R0.1 (Figure 4c), which represents the precipitation frequency, 40% of stations exhibited statistically significant changes. As for SDII (Figure 4b), positive trends were found at 83% of the stations; all stations were less coherent with positive/negative changes. Heavy precipitation, which was represented by R10 (Figure 4a), was dominated by increasing trends. The R0.1 and CWD trends were 1.26 days/decade and 0.047 days/decade, respectively. The SDII trend was weaker. Therefore, the precipitation extremes exhibited significant changes in Xinjiang, specifically regarding the change in rainy days (R0.1), heavy precipitation events (R10) and the mean simple daily intensity index (SDII), indicating that the precipitation changes were reflected in both the frequency and intensity.

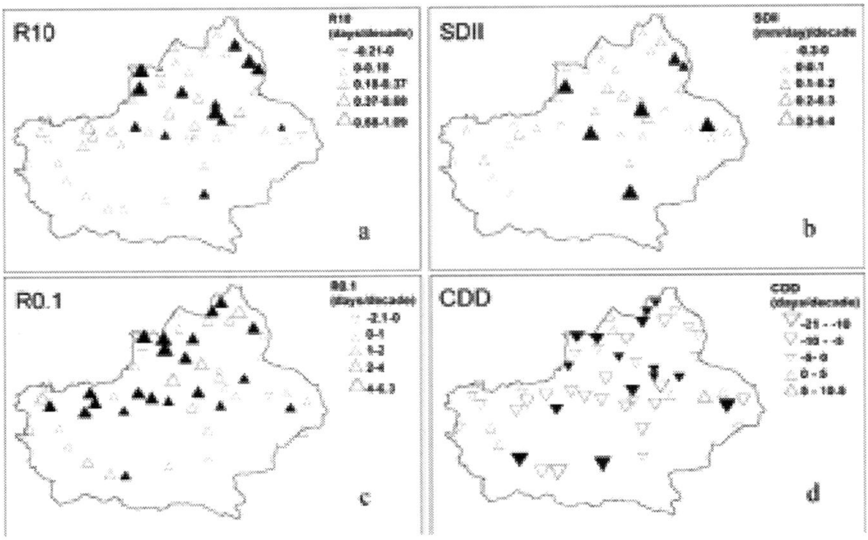

Figure 4: Spatial distribution of the trends and trend magnitudes for precipitation extremes.a: annual R10; b: annual SDII; c: annual R0.1; d: annual CDD. Positive/negative trends are shown as up/down triangles; the filled triangles are related to statistically significant trends (significant at the 0.05 level). The triangle sizes are proportional to the magnitude of the trends.

Characteristics of Climate Extremes

For the annual PDSI in Xinjiang (Figure 5a), 42 stations had significant increasing trends. Stations along the Tianshan Mountains and North Xinjiang exhibited more pronounced trends. The seasonal pattern of the PDSI was same as the annual PDSI (Figure 5b-c), with correlation coefficients exceeding 0.94. A detailed analysis resulted in two contrasting periods (Figure 6). Wet conditions occurred during the period 1987–2010, while persistent drought conditions dominated during the period 1960–1987. The annual PDSI was negative before 1986, and the mean was −1.5, which indicated frequent drought events during these years. Thereafter, the PDSI changed to −0.5; therefore, the PDSI also exhibited a step-wise change around 1986. These results were in accordance with climate (temperature and precipitation) change (Shi et al. [2007]). Moreover, the PDSI exhibited decreasing trends after 2005, indicating the apparent increase in drought severity.

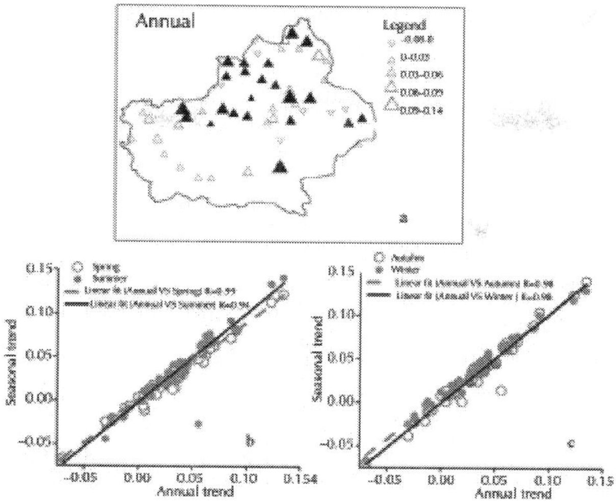

Figure 5: Spatial pattern of decadal trends and spatial characteristic for the seasonal PDIS and the annual PDSI in Xinjiang.a: Spatial trends of annual PDSI and b, c: correlations between the seasonal and annual series.

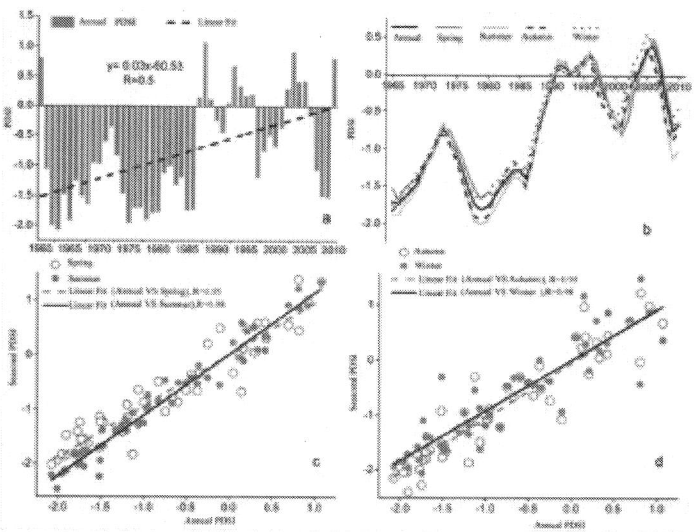

Figure 6: Regional time series of the PDIS in Xinjiang: a: regional time series of the annual PDSI; b: regional five-year moving average of the annual and seasonal PDSIs; and c, d: correlations between the seasonal and annual series.

Influence of Meteorological Disasters on Agriculture

Agriculture is the most sensitive sector to climate change. Affected by climate change, the agricultural growth process is constantly changing. Climate change has a direct effect on agricultural ecosystems, often lead to increased instability and volatility in agricultural production. The crop acreage in Xinjiang increased at a rate of 6.21×10^7 mu/year; the changes can be divided into several distinct phases. The area increased significantly during the periods 1950–1970 and 1995–2012, while the area remained stable during the period 1970–1995 (Figure 7a). The grain crop area decreased substantially after 1970 before reversing in 2008. The area of planted wheat and corn changed in a similar manner to that of the crop acreage, experiencing a progression of increases, decreases, and ultimately increase again. Because of the increased yield per mu, the wheat and corn yields increased substantially (Figure 7c-d). After 2000, the yield/mu maintained a relatively constant level; an increasing trend was not well defined. Therefore, when agriculture developed to a certain level, it was challenging to achieve increasing yields through improved irrigation measures and increased input in the agriculture sector. The economic crop area exhibited a continuous increasing trend. Moreover, in 1998, the planted area exceeded the grain crop area. In the case of cotton, the area, yield and yield per mu increased after 1980 (Figure 7b). Based on the above analysis, the crop acreage increased in Xinjiang primarily due to increased economic crops.

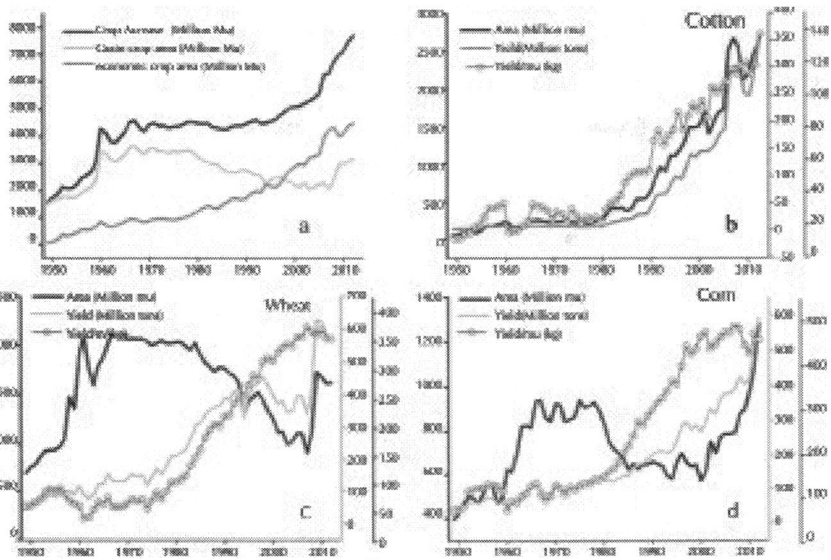

Figure 7: Crop time series in Xinjiang during the period 1949–2012.a: the time series of crop area, and b, c, d: the characteristics of cotton, wheat and corn, respectively.

The statistical results show that the devastated farmland resulting from meteorological disasters increased (either based on the disaster area or the disaster ratio), especially after 1970 (Figure 8a). Flood disaster areas (Figure 8b) were primarily concentrated after 1987, which confirms that the climate in Xinjiang shifted from warm and dry to warm and wet around 1987. Droughts and storms occurred primarily after 1970 (Figure 8c-d). The increasing trend in the storm disaster area was very well defined. For example, the storm disaster area was 210000 acres in the 1960's and increased to 2430000 acres in the 2000's. The proportion affected by low temperatures increased after 1995, especially in 2009, which is when the proportion reached 20% (Figure 8e). This finding indicates that the climate warmed, which simultaneously exacerbated the climate instability and resulted in extreme low temperature events during some periods.

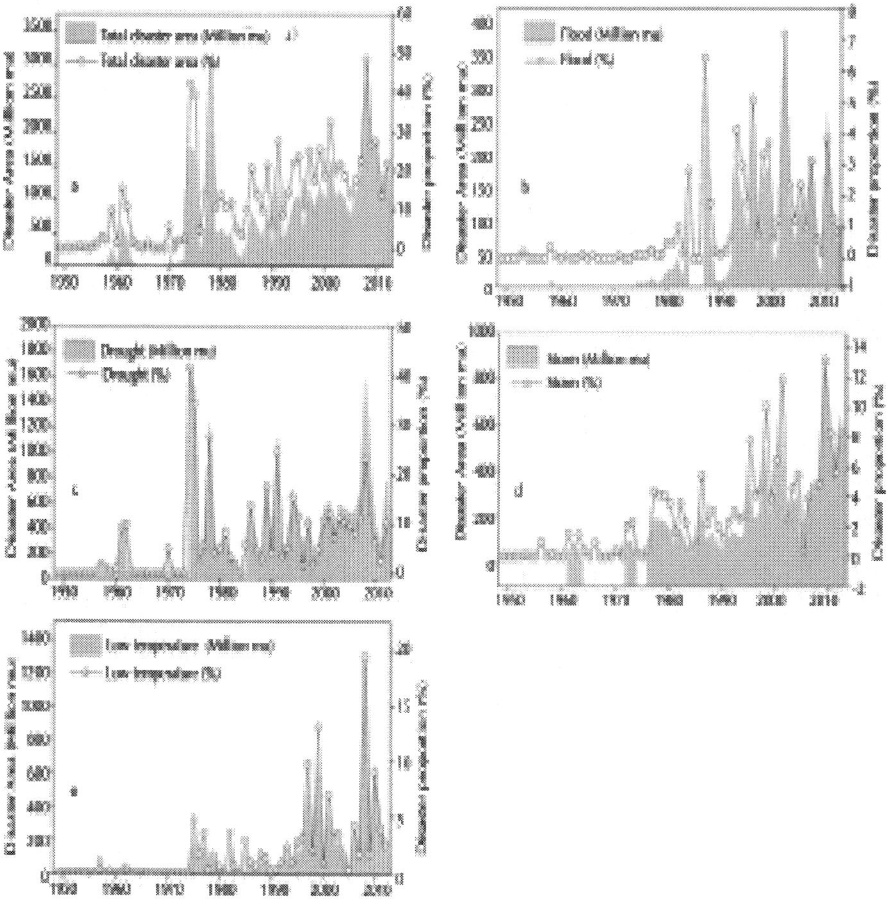

Figure 8: The characteristics of the agricultural disaster area in Xinjiang.a, b, c, d, e is the disaster area and proportion for total disaster area, flood, drought, storm and low temperature, respectively.

The correlation coefficients between the trend in regional climate extremes and the disaster area are shown in Table 2. The statistical data representing the affected areas may be not comprehensive before 1986; therefore, we also discuss data since 1986 via a separate analysis (Table 3). The disaster area exhibited a significant correlation with extreme temperature drought indices, especially the TN10p and TX90p, indicating that the disaster area was primarily controlled by temperature extremes. When the summertime temperature was warm, the corresponding affected farmland area increased. The reasons for floods/droughts are complex and are closely related to the weather, geographical conditions, water conservancy facilities, soil structure, and primarily precipitation extremes. Regardless, changes in the frequency (R0.1), intensity (SDII), amplitude (R10) and duration (CWD) of precipitation events will increase the flood disaster area (Table 2). Drought events exhibited a negative correlation with extreme precipitation events, i.e., when the precipitation intensity and frequency increased, the number of drought events decreased. The enhanced warm temperature extremes (WSDI) resulted in intensified drought. For example, high temperatures in winter increase soil moisture evaporation, affect the safe overwintering of crops, decrease soil entropy, and increase the possibility of soil drought. The PDSI exhibited a negative correlation with drought events, especially after 1986 (Table 3), which is when the correlation was significant. This result indicates that the PDSI partially reflects the drought extent, which can be used to assess drought events in Xinjiang. The storm disaster area was related to the temperature index, indicating that extreme temperatures play a vital role in the formation of strong convective weather. Moreover, the low temperature disaster area was also associated with the extreme temperature index; however, the correlation was found to be positive when compared with warm temperature extremes, which is inconsistent with the actual relationship. This result may be related to climate change variability. For example, within a single year, extremely warm events can occur in combination extremely cold events. The cold events lead to rapid increases in the disaster area. Xinjiang has a typical irrigation agriculture; the shortage of water resources is very serious due to the sharp increase in irrigated area. Based on the exponential growth in rural consumption (Figure 9), we can speculate that the annual extracted groundwater volume increased. Moreover, the effective irrigation area decreased substantially in recent years, which has also resulted in an increase in the disaster area, especially due to the expansion of drought-affected areas.

Table 2: Correlations between climate extremes and the agricultural disaster area in Xinjiang

	Area (Million mu)					Proportion (%)				
	Total	**Flood**	**Drought**	**Storm**	**Low temperature**	**Total**	**Flood**	**Drought**	**Storm**	**Low temperature**
CWD	.057	.581**	-.119	.199	.123	-.018	.551**	-.175	.192	.083
SDII	-.067	.572**	-.223	.081	-.011	-.112	.546**	-.231	.049	-.033
R0.1	.043	.547**	-.105	.249	.068	-.024	.549**	-.179	.261	.008
R10	.051	.738**	-.207	.280*	.148	-.038	.703**	-.267	.277*	.113
RX1day	.057	.656**	-.110	.176	.040	.014	.621**	-.147	.182	.026
CSDI	.005	-.138	.158	-.349*	-.063	.046	-.114	.195	-.390**	-.091
TN10p	-.651**	-.532**	-.356*	-.662**	-.589**	-.518**	-.460**	-.212	-.627**	-.571**
TNn	.101	.207	-.074	.358*	.090	.082	.196	-.066	.408**	.164
DTR	-.380**	-.628**	-.159	-.477**	-.311*	-.299*	-.607**	-.073	-.484**	-.291*
TX90p	.533**	.397**	.276	.492**	.534**	.441**	.335*	.168	.482**	.542**
TXx	.329*	-.070	.277	.209	.377**	.296*	-.106	.222	.205	.376**
WSDI	.470**	.375**	.236	.412**	.551**	.368**	.317*	.116	.419**	.553**
PDSI	.046	.565**	-.088	.147	.060	.013	.561**	-.103	.176	.068

*Indicates significance at <0.05 and **indicates significance at <0.01.

Wu *et al.*

Wu *et al.* *Geoenvironmental Disasters* 2015 **2**:3, doi: 10.1186/s40677-015-0015-8

Table 3: Correlations between climate extremes and the agricultural disaster area in Xinjiang after 1986

	Area (Million mu)					Proportion (%)				
	Total	**Flood**	**Drought**	**Storm**	**Low temperature**	**Total**	**Flood**	**Drought**	**Storm**	**Low temperature**
CWD	-.289	.490*	-.363	-.242	-.146	-.363	.449*	-.339	-.270	-.202
SDII	-.319	.635**	-.527**	-.162	-.186	-.408*	.604**	-.566**	-.163	-.197
R0.1	-.299	.438*	-.423*	-.043	-.248	-.403	.438*	-.448*	-.069	-.313
R10	-.211	.710**	-.542**	.003	-.062	-.327	.654**	-.612**	-.017	-.091
RX1day	-.382	.610**	-.531**	-.213	-.279	-.464*	.568**	-.545**	-.224	-.287
CSDI	.316	.208	.213	-.050	.404*	.283	.268	.115	-.079	.346
TN10p	-.672**	-.177	-.401	-.441*	-.528**	-.536**	-.029	-.184	-.321	-.474*
TNn	-.310	-.206	-.287	.019	-.281	-.303	-.211	-.230	.010	-.210
DTR	-.030	-.410*	.102	-.090	.018	.104	-.348	.198	-.014	.098
TX90p	.544**	.186	.282	.295	.510*	.502*	.091	.132	.253	.517**
TXx	.343	-.174	.193	.163	.436*	.288	-.265	.085	.157	.427*
WSDI	.567**	.120	.406*	.177	.517**	.564**	.026	.293	.158	.520**
PDSI	-.521**	.352	-.461*	-.420*	-.366	-.513*	.364	-.351	-.392	-.374

*Indicates significance at <0.05 and **indicates significance at <0.01.

Wu *et al.*

Wu *et al.* *Geoenvironmental Disasters* 2015 **2**:3, doi: 10.1186/s40677-015-0015-8

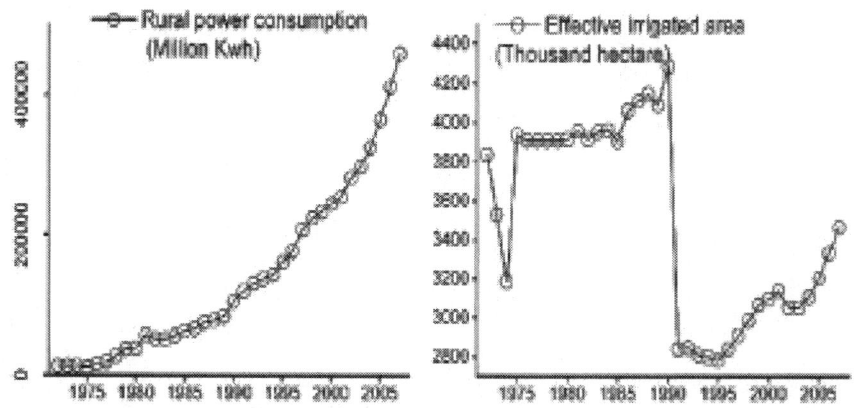

Figure 9: Changes in rural electricity consumption and the effective irrigation area in Xinjiang.

CONCLUSIONS

The characteristics of Xinjiang meteorological disasters were analyzed using extreme indices and the PDSI. Combined with data for the farmland disaster area, we also analyzed the influence of meteorological disasters on the agricultural and ecological systems. The primary conclusions of this study are as follows:

- In Xinjiang, the climate became warmer and wetter, with cold extremes decreasing and warm extremes increasing. Moreover, climate extremes derived from daily minimum temperatures were more numerous that the extremes derived from daily maximum temperatures, which was found to be consistent with the significantly decreased DTR. Precipitation extremes also increased due to the increase in both precipitation frequency and intensity. Frequent and long droughts occurred from 1960 to 1986, although wet climates prevailed from 1987 to 2010. However, after 2003, droughts became more prolific, which was reflected in a drastic decrease in the PDSI.
- Extreme temperatures had a significant effect on the farmland disaster area in Xinjiang. For example, summertime warm events resulted in an increase in the affected farmland area. Floods were

primarily related to extreme precipitation, while the storm disaster area and the low temperature disaster area were the result of changes in temperature extremes. Climate change can also affect the agricultural ecosystem and the oasis crop yield and quality in various ways, leading to increased instability in agricultural production. This study showed that the strengthening of climate warming and evaporation capacity not only increased the soil moisture but also decrease the soil entropy, increased spring drought, accelerated the production of organic matter due to soil microbial decomposition, resulting in the decline in soil fertility and decreased yields. To maintain soil fertility, the fertilization amount must also increase. Although the climate warming can improve grain yields in some regions, climate warming can cause the agriculture water consumption per unit to increase, which also increases production costs. Especially in winter, an increase in extreme minimum temperatures will reduce overwinter human mortality, expand agricultural pest regions, increase the pesticide control difficulty and application amount.

AUTHORS' CONTRIBUTIONS

MW carried out data analysis and write the manuscript. YC provided suggestions and advices to the study and checked the manuscript. HW and GS provided unpublished data and added some useful comments to improve the paper. All authors read and approved the final manuscript.

ACKNOWLEDGEMENTS

The research is supported the Natural Sciences Foundation of China (Grant No.41361093). The authors thank the National Climate Central, China Meteorological Administration, for providing the meteorological data for this study.

REFERENCES

1. Alexandersson H: A homogeneity test applied to precipitation data. *J Climatol* 1986, 6(6):661-675.

2. Bai YG, Musha RZ, Lei XY, Zhang JH: Analysis on characteristics and affecting factor of drought disaster of Xinjiang. *Yellow River* 2012, 34(07):61-63.
3. Buishand TA: Some methods for testing the homogeneity of rainfall records. *J Hydrol* 1982, 58(1–2):11-27.
4. Chen YF, Gao G: An analysis to losses caused by meteorological disasters in China during 1989–2008. *Meteorological Mon* 2010, 36(02):76-80.
5. Chen XJ, Liu XF, Yang XH: Current position of natural disasters in Xinjiang and their control countermeasures. *Territory Nat Resour Stud* 2008, 3(030):64-65.
6. Gao XJ, Zhao ZC, Giorgi F: Changes of extreme events in regional climate simulations over East Asia. *Adv Atmos Sci* 2002, 19:927-942.
7. Hong Y, Adler RF: Predicting global landslide spatiotemporal distribution: integrating landslide susceptibility zoning techniques and real-time satellite rainfall estimates. *Int J Sediment Res* 2008, 23:249-257
8. Kendall MG: *Rank-correlation measures*. Charles Griffin, London; 1975.
9. Liu GC: Discussion on the effects of meteorological disasters on the national economy of Xinjiang. *Arid Zone Res* 1995, 12(03):7-14.
10. Liu L, Hong Y, Bednarczyk CN, Yong B, Shafer MA, Riley R, Hocker JE: Hydro-climatological drought analyses and projections using meteorological and hydrological drought indices: a case study in Blue River Basin, Oklahoma. *Water Resour Manag* 2012, 26(10):2761-2779.
11. Loukas A, Llasat MC, Ulbrich U: "Extreme events induced by weather and climate change: evaluation, forecasting and proactive planning" Preface. *Nat Hazards Earth Syst Sci* 2010, 10(9):1895-1897
12. Mann HB: Non-parametric tests against trend. *Econometrica* 1945, 13:245-259.
13. Palmer WC: *Meteorological drought*. Office of Climatology Research Paper 45, Weather Bureau, Washington, D.C; 1965.

14. Peterson TC, Easterling DR, Karl TR, Groisman P, Nicholls N, Plummer N, Parker D:Homogeneity adjustments of in situ atmospheric climate data: a review. *Int J Climatol* 1998, 18(13):1493-1517.
15. Qiang ZA, Wu TW, Song MH, Ma XB, Cai Y, Liang XY: Arid disaster and advances in arid climate researched over northwest China. *Adv Earth Sci* 2001, 6(01):28-38.
16. Qing DH: China's major meteorological disasters and development trend. *China Emerg Rescue* 2008, 6:4-6.
17. Sen PK (1968) Estimates of the regression coefficient based on Kendall's Tau. J. Am. Stat. Assoc 63:1379-1389
18. Shi YF, Shen YP, Kang E, Li DL, Ding YJ, Zhang GW, Hu RJ: Recent and future climate change in northwest china *Clim Change* 2007, 80:379-39.
19. Sun G, Chen Y, Li W, Pan C, Li J, Yang Y: Intra-annual distribution and decadal change in extreme hydrological events in Xinjiang, Northwestern China. *Nat Hazards* 2014, 70(1):119-133.
20. Wells N, Goddard S, Hayes MJ: A self-calibrating Palmer Drought Severity Index. *J Clim* 2004, 17(12):2335-2351.
21. Xu ZX, Chen YN, Li JY: Impact of climate change on water resources in the Tarim River basin. *Water Resour Manag* 2004, 18:439-458.
22. Xu GH, Mao WY, Lu GY: Xinjiang meteorological weather changes and general statement on the work of disaster prevention and reduction. *Desert Oasis Meteorol* 2008, 2(01):50-54.
23. Ye MQ, Chen BH: Research on natural disaster zoning in Xinjiang *J Nat Disasters* 1996, 5(01):14-21.
24. Yue S, Pilon P, Phinney B, Cavadias G: The influence of autocorrelation on the ability to detect trend in hydrological series. *Hydrol Process* 2002, 16(9):1807-1829.
25. Zhang YC, He WX, Li SK: *Introduction to Chinese agricultural meteorological disasters [M]*. China Meteorological Press, Beijing; 1991.

Chapter 3

The Implications of Post Disaster Recovery for Affordable Housing

Jacqueline McIntosh[1]

[1]School of Architecture and Design, Victoria University of Wellington, Wellington, New Zealand

INTRODUCTION

Disasters, both natural and man-made and involving the massive destruction of habitats, have been with us since the beginning of time. While responses have varied by countries, depending on their population, wealth and culture; and while the magnitude of disaster has also varied, as does the capacity for a country to respond, each country strives to restore its infrastructure to at least its original state, if not better.

At the same time, affordable housing is also a persistent and world-wide imperative. Common strategies to respond to this need include subsidy, reducing quality, simplification and self-help (Katz et al, 2003;Johnson, 2006; Thorns, 2006; FAS 2008a; Gurran et al, 2008). Subsidy, involving agency payment for part or all of the costs of housing (and seeking to recovery) has been challenged by issues relating to internal objectives and external teaming initiatives (Bratt, 2002, FEMA 2008a). Quality has been compromised by issues relating to longevity, generosity, politics and lack of agreement about the minimum acceptable level of service to be provided (Anderson et al, 2003; FEMA 2008b, Lindberg et al, 2010). Simplification refers to the stripping of any superfluous details such as decoration, outdoor spaces and methods of construction and delivery. Self-help refers to the potential for self-building and in some cases self-financing (Harris, 1999; Bredenoord and van Lindert, 2010; HUD, 2007, Leishman, 2012). These policies have been implemented to differing degrees of success, but affordability crisis persists (RAND, 2008b).

The prospect of more frequent and more extreme exceptional weather events, engendered by climate change, further exacerbates the housing affordability crisis. Nowhere is the intersection of these two issues—post disaster recovery and affordable housing—more intense than in the recovery programmes following a disaster such as an extreme weather event. In the face of growing public awareness and concern for global warming and the potential for climate change, the challenges of post-disaster recovery and affordable housing have been intensified with a resurgence of interest in sustainability and green building (Hamin, 2008; Levine et al, 2007; Hayles, 2010). The new imperative for sustainable construction demands that *any* new housing be able to weather and mitigate damage from future extreme weather events. No longer is it sufficient to simply temporarily house victims of a disaster. Post-disaster recovery goals demand that this be accomplished using community processes and that new structures are better and stronger, will mitigate damage from future events and will be culturally appropriate (Berke et al, 1993; Comerio, 1998; Kennedy et al, 2008). All this must be achieved within the constraint of affordability.

This chapter explores the problem of providing affordable and sustainable housing in the context of a post-disaster recovery both in the short term and in the longer term. It focuses specifically on the sustainable construction strategies adopted by various US government agencies in the aftermath of Hurricane Katrina. It begins with a description of the response to the hurricane and the unexpected new housing issues that it created. The chapter then explores the sometimes conflicting mandates of federal and state housing providers as well as the providers of affordable housing within the context of a 'sustainability aware' design community. Finally, it identifies some of the broader implications that can be drawn from this case study.

HURRICANE KATRINA AND THE HOUSING MARKET

Hurricane Katrina devastated the US Gulf Coast in August 2005, inflicting major damage on housing, commercial property and infrastructure (Elliott and Pais, 2006; Nigg et al, 2006; Popkin, et al, 2006;FAS, 2008b). While the US has had longstanding experience coping with hurricanes, Hurricane Katrina was the first to impact the central urban area of a major city. Prior to Katrina, the housing stock in the New Orleans region consisted of a mix of single-family dwellings, multi-dwelling units and mobile homes (Nigg et al, 2006). The majority of these were owner-occupied. Most rental housing was privately owned (as opposed to state-owned) and consisted of both single-family homes and multi-unit dwellings purposely built for rental.

During the hurricane, as could be expected, poorly constructed housing sustained greater damage then did housing of better construction (Elliott and Pais, 2006). It was common to find that lower-income families inhabited the more poorly constructed housing (Peacock et al, 2007). Rental housing was generally more poorly constructed than owner-occupied housing. Despite the widespread extent of the damages to housing stock throughout the region, there were significant differences in the extent and intensity of damage among submarkets (Johnson, 2006). The most severely affected subset was the multifamily rental submarket—almost 80 percent of such units were damaged, and one-third of all multifamily rental units suffered severe or moderate damage. About one-half of the single-family residences (owned and rented) suffered damage, and almost 20 percent of all single-family units were severely or moderately damaged (RAND, 2008a).

Following the hurricane, there was a significant increase in demand for affordable rental housing. While the US had typically relied on market forces to supply the bulk of its rental housing requirements, in this instance, the market did not respond as expected (Popkin, et al, 2006). Instead of rushing to rebuild rental housing, many private investors delayed or indefinitely deferred the decision to rebuild. This is evident in the rates of recovery by housing type (Katz, 2008). The rate of recovery moved more rapidly for single-family dwellings than for multifamily units and it was higher for moderately damaged buildings than for severely damaged units. It was also especially true for most severely damaged properties of uninsured and underinsured homeowners and landlords (for example those with multi-unit rental properties). Landlords with severely damaged buildings faced with "overheated" construction costs and financing shortfalls had the ability to "take their investment money elsewhere," delaying rebuilding until the market cooled off. This was deemed to have been particularly true for the 'mom and pop' landlords who depend heavily on the cash flow from rents (Gulf Renewal, 2007 McCarthy, K and Hanson M, 2008).

In addition to those homeowners seeking to restore their housing, some homeowners were not happy with the directions that their former neighbourhoods were taking in the rebuilding process, so they took the cash from their insurance to rebuild elsewhere, often out of state (Johnson, 2006). As a result, the number of private landlords dropped significantly following the hurricane, the proportion of renters relative to owners increased, the demand for affordable housing increased (also inflated by the influx of low-paid construction workers—many of whom were migrants) and the 'market' showed signs of not recovering fast enough to meet demand, putting added pressure on the state to provide affordable housing—fast (Gulf Renewal 2007, Quarantelli, E. 1995).

Federal Emergency Management Response to Hurricane Katrina

Shortly after Hurricane Katrina struck, the Federal Emergency Management Agency (FEMA), 2008b) provided temporary emergency housing, drawing from their existing inventory of temporary trailers and the purchase of 102,000 additional travel trailers. The FEMA trailers were a mix of new and used small trailers (18.5 m^2), larger travel trailers (37 m^2), and even larger mobile homes (see Figure 1).

Figure 1: FEMA travel trailer types.

Designed for mobility and rapid deployment, the trailers were provided on wheels for ease of movement into trailer parks or onto individual lots. To meet the massive demand, the new trailers were manufactured using the least expensive and most readily available materials and methods, then constructed in haste with little time spent drying out in the factory. This was to backfire later when trailers were found to not only be unsustainable due to their relative expense and short life expectancy, but that they were unhealthy for the occupants (Popkin, et al, 2006).

In addition to the trailers, FEMA had also ordered 25,000 Building America Structural Insulated Panel (BASIP) homes (see Figures 2 and 3). The program for the house design was developed in the 1970s and, like the travel trailers they were designed for temporary shelter, for periods that do not exceed 18 months (Thomas-Rees, 2006). These houses differ from travel trailers both in terms of size and construction. Seeking a more sustainable housing option, the proposed BASIP house design uses prefabricated insulated panels for walls and the roof, resulting in greater energy efficiency and improved durability. Each unit has 3 bedrooms and 2 bathrooms and have been designed for expansion through the joining of a second unit to create a 'double wide.' Other proposed features included special shutters to provide future hurricane protection and solar shading, a retractable awning for solar shading and an additional square area.

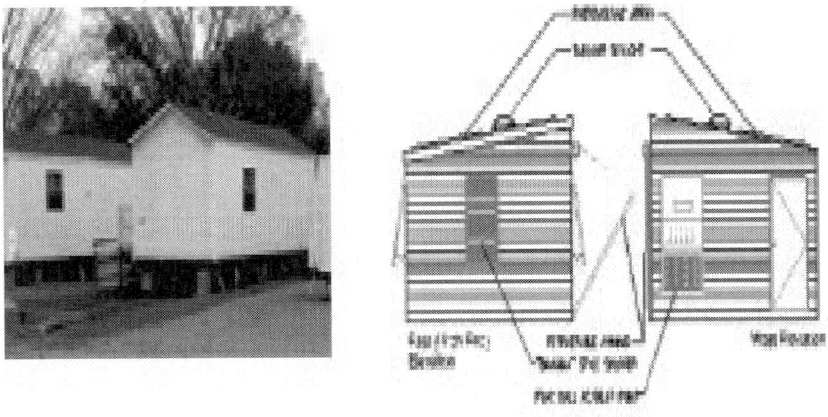

Figure 2: BASIP Homes.

Some of the other sustainable features include the potential for integration of photovoltaics to meet power needs in situations where utilities have not been restored or during times when service is interrupted (Thomas-Rees, 2006). In terms of external appearance however, BASIP's homes look very much like a larger version of the FEMA trailer only with a pitched roof.

Figure 3: BASIPS park in LaPlace, St John Parish, LA.

Overall, the FEMA travel trailers were widely criticized for providing less than desirable temporary housing (Majority Staff Report, 2008). The Internet is filled with personal accounts of unhappy occupants, reports of unhealthy living conditions and overall dissatisfaction. Although the travel trailers were never intended to be used long term (i.e. their use was limited to 18 months), the extended use of travel trailers following disasters of this nature was common in the southern US, with people continuing to live in them for many years. As of mid-August 2007—two years after the hurricane, 60,000 people were still living in 'temporary' shelter FEMA trailers in Louisiana and Mississippi (Blueprint for Gulf Renewal)

State Response to Loss of Housing

While the federal government focussed its efforts on the provision of emergency shelter, the state governments response tended to look further into the future in search of a more holistic and sustainable housing approach that took in the entire urban community (Hassett and Handley, 2006). It became readily apparent to Governor of Mississippi, Hayley Barbour, and his advisors that the so-called temporary shelters had, by necessity, become more permanent fixtures. With their interest in affordable housing models, state officials wanted to go beyond the Band-Aid solutions previously employed for disaster recovery and introduce a more sustainable solution (Peacock et al, 2007). This prompted the governor to employ New Urbanist planner Andres Duany to explore other possibilities. Duany has argued that that shelter is not enough, that a sustainable model had to be fast, flexible and able to transition from temporary shelter to temporary housing and on to permanent housing (Evans-Cowley, 2009). Only through this capacity to transition would costs stay reasonable. Moreover, Peacock et al (2007) point out that the capacity of an individual family household to recover is inextricably tied to housing recovery.

Marianne Cusato

Figure 4: The original Katrina Cottage.

The houses envisaged by Duany and his partner and wife, Elizabeth Plater-Zyberk, were developed as temporary shelter which could be located in new green-field sites, or at the rear of existing properties, as in-fill properties on vacant pockets of land or could even provide additional dwelling spaces for guests or aging relatives. Initial affordability was achieved largely through their small size. Challenging a group of architects to find an alternative to the FEMA trailer, Duany found what he was looking for in a cottage designed by architect Marianne Cusato (see Figure 4). Her house became known as the 'Katrina cottage', named after the hurricane that prompted it's inception. Seeking an intermediate size somewhere between the 18.5 m^2 and 27.8 m^2 FEMA trailers, the first cottage was designed at 27.8 m^2. The cottages were kitset, using prefabricated panels specially designed for hurricane conditions and able to withstand high wind-load conditions and excessive moisture without incurring damage or destruction. In sum, to meet with new objectives, the cottages had to be sustainable, to be able to mitigate damage from future storms, to be appropriate to regional conditions, culture and climate and deliverable by all major delivery methods. This vision extended beyond simple cottage design to an all-encompassing community design, avoiding the less than desirable temporary community plans formerly employed (Talen, 2008; Evans-Cowley, 2009).

Other designers continued to build on Duany and Plater-Zyberk's ideas, expanding the original Katrina cottage idea to 20 different cottage models, including the Kernel House, which was designed to grow from an initial 46.4 m^2 module to a 120.7 m^2 home with added wings (See Figure 5).

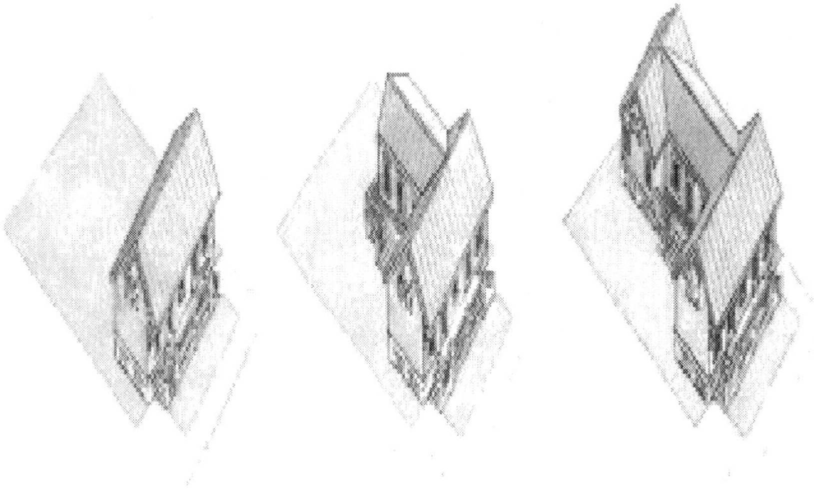

Figure 5: Kernel cottages designed for expansion.

These projects attempted to create sustainable post-disaster recovery housing, but one of the more sustainable and interesting is the Green Mobile Project (see Figure 6), which 'represents a blend of key emergency housing needs with energy efficient and affordable housing that can serve as a temporary or permanent dwelling--emphasizing innovative site design features, green building technologies, durability, expandability with an open interior design that can be adapted to varied family needs. The units, therefore, result in 'reduced energy consumption and affordable living' (United States Government Accountability Office Washington, DC 2007, FEMA website).

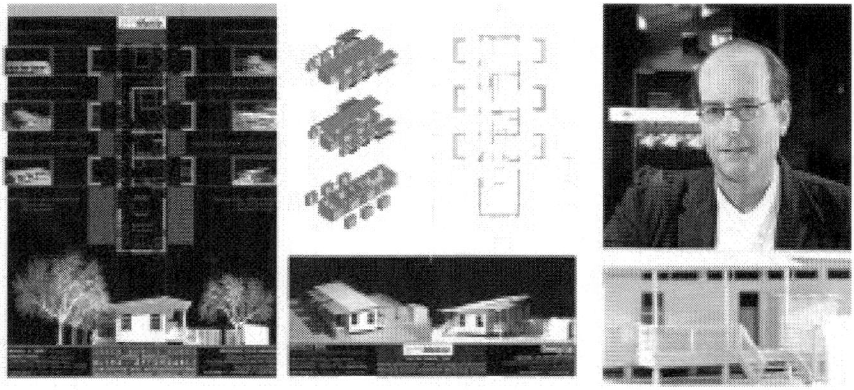

Figure 6: MSU architecture professor Michael Berk shows a model of Green-Mobile.

Post-Disaster Recovery Housing Becomes Affordable (And Desirable)

Meanwhile the prototype Katrina cottage had caught the imagination of designers throughout the southern US, eliciting a host of Katrina copies with varying degrees of sustainable construction. Currently there are dozens of listings on the Internet for Katrina cottages that available for rent (Lane, 2008, Benfield, 2010). They are being used for long-term housing and for uses including vacation homes, granny cottages and home offices (Benfield 2010, Stark, 2006). The trailers arrived on site ready to inhabit, flexible for use in a variety of situations and as temporary dwellings, they did not require building consent. The following quotes from the Internet illustrate the range of uses and locations that the cottages have been deployed, well beyond their intended post-disaster recovery housing based in the New Orleans region.

"The Katrina cottage - with living quarters about the size of a McMansion bathroom - is now appealing to people well beyond the flood plain. Californians want to build one in their backyards to use for rental income to help with the mortgage payment. Modestly paid kayakers in Colorado see it as a way to finally afford a house. Elsewhere, people envision building one so a parent can live nearby." (M Cusato, Katrina cottage designer)

"We have lot sizes that are too small for a... single-family, detached household, so the idea is to bring in these extremely attractive dwellings to provide affordable housing," (City councillor, Connecticut)

'Meanwhile, in the iconic new urbanist resort town of Seaside, Florida, Katrina Cottages are being employed for a small educational facility' (Kaid Benfield)

"A developer in Virginia wants to do some as affordable houses. People see it and realize it's a dignified way to live." (M Cusato)

Driven by the desires of the public, the US had found a desirable housing model suited not only to post-disaster recovery, but to the wider development of sustainable, community-oriented, affordable housing.

CONCLUSIONS AND DISCUSSION

The Hurricane Katrina experience has graphically shown the critical importance of planning and designing post-disaster recovery housing well before the disaster strikes. This advanced planning and design needs to be conducted at both the level of the individual dwelling as well as the community in which that dwelling will be located. It is also critical that these advanced designs be tested particularly when trialling new technologies and processes (Miller, 2006). It is also vital that leadership for these design initiatives be initiated by both the federal and the state government (Bathurst, et al 2011). Once these are initiated, it is imperative that communities and local businesses become fully engaged in the evolutionary design processes *prior* to the disaster (Popkin, 2006; Cowley-Evans, 2009; Talen, 2008).

In planning and designing post disaster recovery housing, it is important to work backwards, from a sustainable housing model that is suited to the needs of the current rental market, but adaptable for disaster situations. It is also critical to develop the manufacturing technologies and delivery processes that can be ramped up in large numbers and in short time frames. These are best resolved well before any emergency arises. This is the only way to get beyond the constraints imposed by supply and demand economics.

The experience of the post-disaster recovery period in the US following Hurricane Katrina suggests that it is possible to build sustainably and affordably, but it is difficult (Johnson, 2007). The following lessons regarding sustainability and affordability can be gleaned from the events that took place in the aftermath of Hurricane Katrina:

- There is no such thing as 'temporary' with respect to housing, post-disaster recovery or otherwise. Sustainable housing is more than a discrete 'product'. It must also be considered as a cultural artefact that has particular meaning for the community within which it is located.
- Significant cost and waste results from non-cooperation between agencies. Conflicting organisational objectives must be aligned in order to achieve long-term sustainable solutions.
- Even in the most liberal capitalist society, the government should endeavour to lead the way by providing well-researched advanced design solutions. The government needs to play an important role modelling function in facilitating a desirable outcome that will allow the market to follow.

The convergence of post-disaster recovery and sustainable housing research practices has fortuitously created important new directions and opportunities. While the solutions that were created in response to Hurricane Katrina might be vastly different from those appropriate for Third World countries or other cultural settings, the process for 'getting to sustainable, affordable housing' is more universal in its application. *Any* new housing must, within the confines of what is affordable, be able to weather and mitigate damage from future extreme weather events, be culturally appropriate, and strengthen community structures. This leaves us with only one major remaining challenge: Who is deserving of sustainable affordable housing? One intriguing possibility has been put forward by a Californian architect from one of the oldest architectural firms based in that state:

'One of the problems that I see with it, and I probably shouldn't say this, is that it looks nice. I think the government has a very hard time giving things away to people or underwriting things that go beyond some sort of bureaucratically understood minimal gesture'.

The political reality that this architect alludes to is that a significant proportion of the wealthier sectors of society are resistant and vocal about granting the very poor considerably better housing stock to that prior to what they had (or perceived to be had) prior to the disaster. In a post-disaster recovery situation this political and socio-cultural hurdle is mitigated by a combination of public sympathy to widespread loss and the benchmark of the pre-existing housing. Overall, replacement housing should be of equal or only marginally better 'value' than that which existed before. In the cases of the extreme poor, the homeless, or those without property, establishing these benchmarks seem to pose an insurmountable problem.

REFERENCES

1. Anderson, St. Charles, J, Fullilove, M, Scrimshaw, S, Fielding, J, Normand, J (2003) 'Providing Affordable Family Housing and Reducing Residential Segregation by Income : A Systematic Review' *American Journal of Preventive Medicine*, Volume 24, Number 3
2. Bathurst, R, Jackson, B and Statler M (2010) 'Leading aesthetically in uncertain times' *Leadership* 6(3) 311–330
3. Benfield, K (2011) "Katrina Cottages" find post-Katrina uses as affordable housing, educational facilities, offices, & more" *Switchboard* posted July 11, 2011
4. Berke, P, Kartez, J, and Wenger D (1993) 'Recovery after Disaster: Achieving Sustainable Development, Mitigation and Equity' *Disasters* Volume 17, No 2, 1993
5. Blueprint for Gulf Renewal: The Katrina Crisis and a Community Agenda for Action. A Special Report by Gulf Coast Reconstruction Watch--A project of the Institute for Southern Studies. August/ September (2007)
6. Bratt, R. G. (2002) 'Housing and Family Well-being' *Housing Studies* Volume 17, No 1, 13-26
7. Bredenoord, J, Van Lindert P and Smets P (2010) 'Equal Access to Housing' *Habitat International* Volume 34, Issue 3, July, Pages 278–287
8. Comerio, Mary C. (1998) *Disaster Hits Home: New Policy for Urban Housing Recovery*, Berkeley: University of California Press
9. Elliott J, Pais, J (2006) 'Race, class, and Hurricane Katrina: Social differences in human responses to disaster' *Social Science Research* 35 295–321

10. Evans-Cowley, J, Zimmerman Gough, M (2009) 'Evaluating New Urbanist Plans in Post-Katrina Mississippi, *Journal of Urban Design*, Vol. 14. No. 4, 439–461, November
11. Federation of American Scientists (FAS) (2008a), Two & a Half Years Later: Surviving the FEMA Aftermath: http://www.fas.org/programs/energy/btech/about/Surviving%20the%20FEMA%20Aftermath.pdf
12. Federation of American Scientists (FAS) (2008b), A Response to Katrina: The making of the largest, most efficient modular building order in US history http://www.fas.org/programs/energy/btech/about/MEMA%20AHPP%20Article.pdf
13. FEMA (2008a) Katrina/Rita Housing Facts: HQ-08-008Factsheet http://www.fema.gov/news/newsrelease.fema?id=42417
14. FEMA and CDC Joint Press Release (2008b), FEMA 'To Expedite Relocation Of Residents From Temporary Housing Units' http://www.cdc.gov/media/pressrel/2008/r080214b.htm
15. FEMA technical notes: Alternative Housing Pilot Program (AHPP) and Joint Housing Solutions Group (JHSG) (nd) posted at http://www.huduser.org/periodicals/cityscape/technotes/AHPP.pdf and http://www.huduser.org/periodicals/cityscape/technotes/JHSG.pdf.
16. Gurran, N, Milligan, V, Baker, D, Beth Bugg L. and Christensen, S (2008) 'New directions in planning for affordable housing: Australian and international evidence and implications' Australian Housing and Urban Research Institute, Sydney Research Centre June 2008 AHURI Final Report No. 120
17. Hamin E.M., & Gurran, N. (2008) 'Urban form and climate change: Balancing adaptation and mitigation in the U.S. and Australia', *Habitat International* , doi:10.1016/j.habitatint.2008.10.005
18. Harris, R (1999) 'Slipping through the Cracks: The Origins of Aided Self-help Housing, 1918-53' *Housing Studies* Volume 14, Issue 3, 281-309
19. Hassett, W. Handley, M (2006) 'Hurricane Katrina Mississippi's Response' Public Works Management Policy April vol. 10 no. 4 295-305
20. Hayles, C. (2010) "An examination of decision making in post disaster housing reconstruction", *International Journal of Disaster Resilience in the Built Environment*, Vol. 1 Iss: 1, pp.103 - 122

21. Lindberg, R. Shenassa, E, Acevedo-Garcia, D. Popkin, S. Villaveces, A., Morley R (2010) 'Housing Interventions at the Neighborhood Level and Health: A Review of the Evidence', *Journal of Public Health Management Practice*, 16(5) E-Supp. S44–S52

22. Johnson, C (2007) 'Strategic planning for post-disaster temporary housing' *Disasters*, 31(4): 435–458.

23. Johnson, M. (2006) 'Decision Models for Affordable Housing and Sustainable Community Development' *Journal of the American Planning Association*: The Future(s) of Housing

24. Katz, C (2008) 'Bad elements: Katrina and the scoured landscape of social reproduction' *Gender, Place and Culture*, Vol 15, No1. February 2008, 15-29

25. Katz, B, Austin, M. Turner, K. Brown, D.. Cunningham, M, Sawyer. N (2003) 'Rethinking Local Affordable Housing Strategies: Lessons from 70 Years of Policy and Practice: A Discussion Paper prepared by The Brookings Institution Center on Urban and Metropolitan Policy and The Urban Institute December 2003

26. Kennedy, J, Ashmore, J, Babister, E and Kelman, I (2008) 'The Meaning of 'Build Back Better: Evidence From Post-Tsunami Aceh and Sri Lanka'*Journal of Contingencies and Crisis Management* Volume 16 Number 1 March

27. Lane, M (2008) 'A Cottage Industry: Homes created for Hurricane Katrina victims may find a home here' The St. Augustine Record. St Augustine 01/27/08 Date of access 19/11/2012 Available from http://staugustine.com/stories/012708/news_txt01_26.shtml

28. Leishman C.and Rowley, S. (2012) ' Pro-poor housing policies: Rethinking the potential of assisted self-helphousing' *Affordable Housing* , in The SAGE Handbook of Housing Studies edited by William A V Clark, Kenneth Gibb, David F Clapham pp 379-396

29. Levine, J. Esnard, A and Sapat, A (2007) 'Population Displacement and Housing Dilemmas Due to Catastrophic Disasters' *Journal of Planning Literature* August 2007 vol. 22 no. 1 3-15

30. Majority Staff Report (2008) 'Toxic trailers - Toxic Lethargy: How the Centers for Disease Control and Prevention has Failed to Protect the Public Health' Subcommittee on Investigations & Oversight, Committee on Science and Technology, US House

of Representatives. Available from http://oversight.house.gov/documents/20071114164004.pdf

31. Miller, J (2006) 'Katrina cottages, one year later' The Town Paper Available from http://www.tndtownpaper.com/Volume8/katrina_cottages.htm
32. McCarthy, K and Hanson M (2008) Post-Katrina Recovery of the Housing Market Along the Mississippi Gulf Coast, Technical Report of the Gulf States Policy Institute, RAND Corporation.
33. Nigg, J. Barnshaw, J. Torres, M. (2006) 'Hurricane Katrina and the Flooding of New Orleans: Emergent Issues in Sheltering and Temporary Housing' *The ANNALS of the American Academy of Political and Social Science* March Vol. 604 no. 1 113-128
34. Peacock, W.G., Dash, N, Zhang, Y (2007) 'Sheltering and Housing Recovery Following Disaster' *Handbook of Disaster Research, Handbooks of Sociology and Social Research* pp 258-274
35. Popkin, S, Turner, M and Burt, M (2006) 'Rebuilding Affordable Housing in New Orleans: The Challenge of Creating Inclusive Communities' After Katrina Rebuilding Opportunity and Equity into the New New Orleans The Urban Institute January 2006
36. HUD (2007) Planning for Catastrophe (Symposium), US Department of Housing and Urban Development, Office of Policy development and Research. Published as Cityscape cityscapevol9num3 (2008) http://www.huduser.org/periodicals/cityscpe/vol9num3/cityscapevol9num3.pdf
37. Quarantelli, E. (1995),"Patterns of sheltering and housing in US disasters", *Disaster Prevention and Management*, Vol. 4 Iss: 3 pp. 43 – 53
38. RAND Corporation report (2008a) Rebuilding Affordable Housing on the Gulf Coast Lessons Learned in Mississippi from Hurricane Katrina
39. RAND Corporation report (2008b) Post-Katrina Recovery of the Housing Market Along the Mississippi Gulf Coast, Technical Report of the Gulf States Policy Institute, http://www.rand.org/pubs/technical_reports/2007/RAND_TR511.pdf
40. United States Government Accountability Office Washington, DC (2007) Subject: Disaster Housing: Implementation of FEMA's Alternative Housing Pilot Program Provides Lessons for Improving

Future Competitions; GAO-07-1143R http://www.gao.gov/new.items/d071143r.pdf

41. Stark, J (2006) 'Katrina cottages' get touch of Florida' Tampa Bay Times, Published August 26, 2006 Available from http://www.saintpetersburgtimes.com/2006/08/26/Homes/_Katrina_cottages__ge.shtml
42. Talen, E (2008) 'New Urbanism, Social Equity, and the Challenge of Post-Katrina Rebuilding in Mississippi' *Journal of Planning Education and Research* 2008 27: 277
43. Thomas-Rees, S., Chandra, S., Barkazsi, S., Chasar, D., Colon, C., "Improved Specificiations for Federally Procured Ruggedized Manufactured Homes for Disaster Relief in Hot/Humid Climates". Paper published in the Proceedings of the 15th Symposium on Improving Building Systems in Hot and Humid Climates. July 24-26, 2006.
44. Thorns, D (2006) 'The Remaking of Housing Policy: The New Zealand Housing Strategy for the 21st Century' *Housing Finance International* – June 2006
45. Katrina cottages website: http://www.katrinacottages.com

Chapter 4

Maternal Anxiety 16 Months after the Great East Japan Earthquake Disaster Area: First Report

Hatsumi Yoshii[1], Hidemitsu Saito[1], Saya Kikuchi[2], Takashi Ueno[3], and Kineko Sato[1]

[1]Health Sciences, Tohoku University, Graduate School of Medicine, Miyagi, Japan

[2]Department of Psychiatry, Tohoku University Hospital, Sendai, Japan

[3]Division of Clinical Psychology, Graduate School of Education, Tohoku University, Sendai, Japan

ABSTRACT

The Great East Japan Earthquake, a magnitude 9.0 quake that occurred on March 11, 2011, left more than 20,000 killed or missing and resulted in more than 400,000 people being displaced. The Fukushima Nuclear

Power Plant accident released large amounts of radioactive material into the air. Among the victims of this combined disaster were many pregnant and parturient women, and this study aimed to determine post-disaster anxiety among this specific population and mea- sures for the future. Participants were 259 women (mean age 33.02 ± 4.79 years) who gave birth around the time of the earthquake in Miyagi Prefecture, one of the disaster areas. Sixteen months after the earthquake, we administered survey questionnaires on anxiety. We transcribed questionnaire responses, coded raw data by context, and categorized these codes by commonality. After extracting subcategories of anxiety-related factors, we categorized them into more abstract concepts. Among the participants, 126 (48.6%) reported having no available professionals with whom they could consult about childrearing. Participants reported anxiety in the following 12 categories: "radiation," "child's physical and mental growth/development," "recurrence of earthquake and tsunami," "financial issues," "childrearing environment," "living environment," "maternal employment," "stigma," "familial issues," "maternal health," "childrearing," and "the future". A beneficial way to reduce maternal anxiety in the 12 areas identified would be to develop support systems that provide continuous support for children's mental health care needs, psychological guidance, community support for maternal empowerment, outreach for individual support, and professional consultation for mothers who have high anxiety about radioactivity.

INTRODUCTION

The Great East Japan Earthquake, a magnitude 9.0 quake that occurred on March 11, 2011, was centered offshore of Sanriku in Miyagi Prefecture, northeastern Japan [1] . Miyagi, Iwate, and Fukushima prefectures were struck by the earthquake and subsequent tsunami. As of July 31, 2013, over 20,000 were missing or killed (Figure 1), and over 400,000 people were evacuated in the aftermath [2] . Many people in the areas flooded by the tsunami had to live in shelters for several months, while some moved inland [3] . In addition, the Fukushima Nuclear Power Plant accident caused by the tsunami released large amounts of radioactive materials into the air [4] . On March 12, 2011, the government issued an evacuation order for people living within a

20-km radius of the plant [5] . Many pregnant and parturient women lived in these areas. It is well documented that women in general are particularly susceptible to adverse effects of a disaster [6] - [10] and the March 11 tragedy reawakened our awareness of the need to manage maternal anxiety after the acute phase of a disaster.

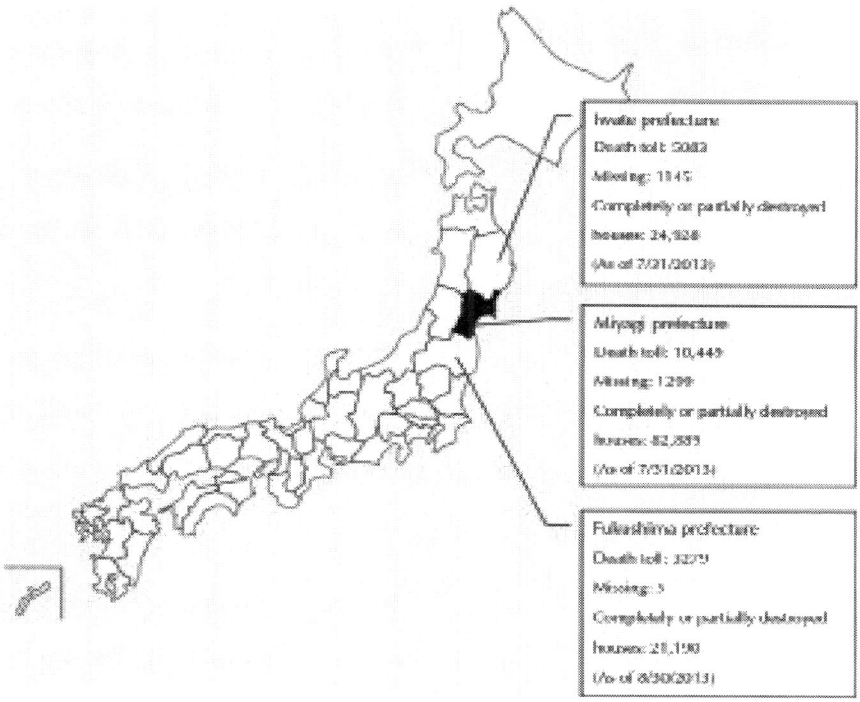

Figure 1: Disaster Areas following the Great East Japan Earthquake, with Miyagi Prefecture shown in Black from which the Participants Were Recruited. Sources: Fukushima Prefecture Earthquake Damage Information http://wwwcms.pref.fukushima.jp/pcp_portal/PortalServlet?DISPLAY_ID=DIRECT&NEXT_DISPLAY_ID=U000004&CONTENTS_ID=24914; Iwate Prefecture Disaster Prevention Portal Site http://www.pref.iwate.jp/~bousai/; Miyagi Prefecture Official Website http://www.pref.miyagi.jp/site/ej-earthquake/.

This study reports the results of a survey on post-disaster anxiety administered 16 months after the Great East Japan Earthquake to mothers in Miyagi Prefecture who had delivered a child around the time of the earthquake in order to investigate various sources of anxiety in their lives and determine the most appropriate means of support.

METHODS

Subjects

Study participants were mothers who had delivered a child less than one month before the Great East Japan Earthquake that occurred on March 11, 2011, and pregnant women who had received a "Maternal and Child Health Handbook," the handbook issued by Japanese municipal governments in accordance with Article 16 of the Maternal and Child Health Law for pregnant women who report their pregnancy to a municipal office. We contacted 26 medical institutions in Miyagi Prefecture and asked them to explain the purpose of this study to their patients who gave birth during the period from February 2011 through September 2011. We administered the survey in July 2012.

Ethical Considerations

Approval to conduct this research was obtained from the Graduate School of Medicine, Tohoku University and hospitals and government offices in Tohoku, Japan that agreed to participate after receiving an explanation of the study. We informed the participants that the content of the investigation would be used for research only and that they could withdraw from the study at any time. We also explained that their participation would be kept confidential. Before starting the investigation, we obtained signed informed consent forms from all participants.

Questionnaire Survey

The survey questionnaire included open-ended questions asking participants to describe their anxieties. After collecting basic demographic information, the questionnaire began with the general question, "What is causing you anxiety?" to which participants could freely express their responses.

Analytical Methods

We transcribed the questionnaire responses, coded anxiety-related raw data by context, and categorized these codes by commonality. We then extracted subcategories of these anxiety-related factors and further categorized them into more abstract concepts. We used ATLAS. ti. 7.0 software (Scientific Software Development GmbH, Berlin) for data analysis. We chose this analytical method because it feeds appropriately into a concept analysis rather than a qualitative theory [11] . Data analysis was started with iterative reading to facilitate familiarity with the data, allowing for the beginnings of an interpretative process. To enhance credibility of the analytical process, the co-authors analyzed the data, verified the coding, and organized the data into themes.

RESULTS

Participant Characteristics

Among 886 women who initially agreed to participate, the final sample comprised 259 women who returned their questionnaires. The mean age of participants was 33.02 ± 4.79 years. Most participants (179, 69.1%) were in their 30 s. Among all participants, 99 (38.2%) were primiparae and 160 (61.8%) were multiparae. In addition, 49 (18.9%) had to move to another residence after the earthquake, most of whom (17, 34.7%) moved to rental housing provided by the prefecture followed by 6 (12.2%) who moved to temporary housing. As for their general health condition since the earthquake to the present time, 179 (69.9%) reported they were "healthy," 61 (23.8%) reported that they "became ill, but recovered," and 16 (6.3%) reported that they were "still ill." In addition, 126 (48.6%) reported the unavailability of childrearing professionals (e.g., physicians or public health nurses) with whom they could consult (Table 1).

Maternal Anxiety

We classified 454 codes for maternal anxiety into 66 subcategories. We then classified these subcategories into 12 higher level categories:

"radiation," "child's physical and mental growth/development," "recurrence of earth- quake and tsunami," "financial issues," "childrearing environment," "living environment," "maternal employment," "stigma," "familial issues," "maternal health," "childrearing," and "the future." Table 2 describes each category.

Table 1: Demographic characteristics of participants

Characteristic	n	%
Current residence		
No change	210	81.1
Temporary housing	6	2.3
Rental housing	17	6.6
Own parent's house	5	1.9
Husband's parents' house	2	1
Other	19	7.1
Employment		
Yes	107	41.3
No	152	58.7
Type of employment		
Side job	2	0.8
Part time	25	9.7
Full time	57	22
Self-employed	7	2.7
On maternity leave	5	1.9
Other	11	4.2
None	152	58.7
Health condition after the earthquake		
Healthy	179	69.1
Became ill, but recovered	61	23.6
Still ill	16	6.2
No answer	3	1.2
No. of children		
1	99	38.2
2	110	42.5

3	39	15.1
4	5	1.9
5	4	1.5
6	1	0.4
No answer	1	0.4
Currently pregnant		
Yes	14	5.4
No	245	94.6
Have someone I can consult about children's issues		
Definitely yes	169	65.3
Somewhat yes	72	27.8
Somewhat no	13	5
Definitely no	4	1.5
No answer	1	0.4
Have someone I can consult about childrearing (e.g., professionals such as physician and public health nurse)		
Definitely yes	59	22.8
Somewhat yes	73	28.2
Somewhat no	71	27.4
Definitely no	55	21.2
No answer	1	0.4

Table 2: Mother's anxiety in a disaster area

Category	Subcategory		
Radiation	Food safety		
	Outdoor safety		
	Influence on an unborn child		
	Influence on children		
	Influence of radiation on the human body		

	Financial damage		
	Distrust of public announcements		
Child's physical and mental growth/development	Mental impact of disaster experience		
	Mental impact of parents' instability		
	Developmental delay and issues		
	Illness		
	Future health condition		
	Weight loss or gain		
Recurrence of earthquake and tsunami, and aftershocks	Ability to protect children		
	Evacuation		
	Recurrence		
	Anxiety during the time apart from one's own child		
	Anxiety during husband's absence		
	Lifelines		
Financial issues	Housing loan issues caused by the earthquake		
	Financial damage caused by the earthquake		
	Money for the future		

	Living expenses		
	Reserve for child's education		
	Consumption tax increase		
	Low income due to a single parent household		
Childrearing environment	Playground		
	New neighborhood		
	Daycare center		
	Child's making friends		
	Mother's making friends		
	Shortage of pediatricians		
	Future life after childbirth		
	New school		
Living environment	Unable to find permanent housing		
	Remaining in a disaster area		
	Life in temporary housing		
	Reconstruction		
	Crimes		
Maternal employment	Returning to work		
	Daycare for children while working		
	Balance between work and childrearing		

	Child's becoming ill		
	Job search		
	Responsibilities at the workplace		
	Sense of guilt over sending child to daycare center		
	Sense of guilt about maternity leave		
Stigma	Being stigmatized as a disaster victim		
	Being stigmatized by other disaster victims		
Familial issues	Health condition of parents and relatives		
	Relationship with parents and relatives		
	Husband's employment and health condition		
	Relationship with husband		
Maternal health	Fatigue		
	Poor conditions		
	Mental issues		
	No physical energy		
	Currently pregnant		
Childrearing	No available professionals to consult		
	Discipline		

	How to communicate with children		
	Weaning		
	Meals		
	Sleep		
Future	Future life		
	Future of Japan		

DISCUSSION

The total number of fatalities in Miyagi prefecture was 10,449 (Figure 1). Strong aftershocks struck the area daily for at least the first month [12] . One month after the earthquake, there was a strong aftershock with seismic intensity in the high sixes in Miyagi Prefecture, then in December 2012, another tsunami warning was issued and a 1-meter high tsunami was observed in Ishinomaki City. Aftershocks still occur intermittently, and they have kept people in these areas alert and anxious even 2 years later. Given this situation, study participants reported strong anxiety about the recurrence of earthquakes and tsunamis along with aftershocks. Specifically, they were concerned about "whether they are able to protect their children," and "whether they are able to guide their children safely to evacuation areas." Japanese government agencies and the mass media still provide information about evacuation routes and survival kits for disasters. In the prefecture, 82,889 houses were completely destroyed (Figure 1). Approximately 2853 victims were evacuated to shelters within the prefecture and 8524 were evacuated outside it. Even 16 months later, many respondents were still living in temporary housing with no plans for a permanent residence or living within the disaster area, which led them report "anxiety about their living environment." Accordingly, their "childrearing environment" also changed, which led to related anxieties including adjustment to new locations and facilities, as well as the continued contamination of many playgrounds and a resulting shortage of safe play areas for children. Respondents also reported anxiety related to "family issues" and "stigma" because many of their family members (parents, siblings, and other relatives) were also victims. Among those who reported anxiety about "financial issues," some worried about housing loan issues due to the earthquake. Although their anxiety about "childrearing," "maternal employment," and "the future" might be common to all mothers, the physical and mental health condition of these victims is far more serious than those living under normal conditions, as evident from the following: 1) more than 30% of respondents had physical problems after the earthquake; 2) they often reported "anxiety about their own health"; 3) many respondents scored below the median for "physical condition" (46.9%) and "anxiety and sleeplessness" (92.4%); and 4) results on the General Health Questionnaire 28 administered by one of the authors (K.S.) to the same sample [13] revealed that scores (>0.6) of more than 66.4% of these respondents denoted a high risk of developing mental health problems. Disasters occur unexpectedly and suddenly, thereby forcing people to face the fear of death, or actual death, which often induces posttraumatic stress reaction in survivors, some of whom are subsequently diagnosed with posttraumatic stress disorder (PTSD) [14] . In many cases of PTSD, a wide range of difficulties often coincide, such as anxiety disorders, mood disorders, and drug and alcohol abuse. In addition to traumatic stress, disasters often induce other types of stress, including sense of loss and guilt. Therefore, a thorough investigation of disaster victims' mental condition, including PTSD, is needed, and it is equally necessary to provide them every possible means of support [15] .

Respondents also expressed anxiety about their "child's physical and mental growth/development". Catastrophic disasters such as earthquakes and war cause tremendous physical and mental damage [16]. McLaughlin and colleagues interviewed 797 children (aged 4 - 17 years) 27 months after a hurricane and found 15% were experiencing serious emotional disturbances, including aggression, self-injury, excessive fear, anxiety, withdrawal, learning difficulties, and PTSD. After 36 - 39 months, 11% continued to experience these problems [17] . PTSD is an anxiety disorder that occurs in the aftermath of a traumatic event [16] . Similarly, mothers in this study also felt insecure, and the following codes emerged: "being incapable of managing mental health care in sensitive and fearful conditions," and "having no idea what to do with children's 'tsunami game' play." The tsunami game is a type of posttraumatic play in which traumatized children repeatedly reproduce the index event. These phenomena appeared to be improved 1 year after the earthquake. However, some children still exhibited PTSD symptoms, and others sometimes showed these symptoms long after the earthquake, causing trouble for some schools and daycare centers to handle, including truancy, reluctance to attend school, bullying based on discrimination of victims, and impulsive behaviors such as wrist cutting [18] . As part of the "Mental Health Care Team" activities sponsored by the Japanese Ministry of Health, Labour and Welfare, Umezaki and Arachi explained to mothers that these phenomena are temporary acute stress disorders that are typical for children in disaster areas, and they counseled them about how to communicate with their children by providing brochures (i.e., Handbook for Families with Children Who Experienced a Disaster [The Japanese Society for Child and Adolescent Psychiatry], Children Who Experienced a Disaster—Mental Health Care, and For Mental Health Care [Miyagi Prefecture Mental Health Care Center]) [19] . As Jones and Schmidt reported, children of different ages and developmental stages are affected differently by disaster, therefore the impact of disasters on children does not always immediately appear [17] . In fact, they can resurface and persist over a long period of time. Consequently, long-term support systems that provide access to consult services on childrearing are necessary for the maintenance and promotion of both physical and mental health.

Another characteristic finding of this study is anxiety about radiation, which is equal to the fear of earthquakes or tsunamis. Miyagi Prefecture, where our hospital is located, is in eastern Japan, ~68 km (~42.5 miles) southeast of the Fukushima nuclear power plant (Geospatial Information Authority of Japan). While the prefecture was not designated a mandatory evacuation zone, many mothers with young children were concerned about radioactive contamination of food and playgrounds. The prefectural government assessed radiation in agricultural and fishery products, and quarantined any food that contained radiation levels with higher than normal background values. They also assessed radiation levels of playground soil and deployed decontamination procedures on ground where radiation levels were abnormal. Nevertheless, this study showed that mothers were not completely reassured by these efforts.

A wide range of support systems should be provided to reduce maternal post-disaster anxiety. For example, continuous support programs in local communities are needed such as the "Mental-Health-

Care Team for Children," "Children's Mental Health Counseling," and "Children's Mental Health Counseling Health Checkup," all of which were provided by other areas outside the disaster zones. Also, accurate information should be provided about the physical and mental development of children who have lived through disasters. This could entail psychological guidance or counseling mothers to encourage and empower each other. Support providers should not be limited only to professionals such as clinical psychologists, public health nurses, midwives, and nurses, but ideally would also include radiation specialists who can help those experiencing high levels of anxiety about radiation, as revealed in this study. It would appear feasible to provide periodic local support programs for parents with preschoolers in facilities such as daycare centers, kindergartens, and local child centers because these are sites that host periodic health check-up clinics. Also, we can address the issue of the unavailability of professionals with whom mothers can consult about childrearing, as reported by over half of this study's participants, by establishing outreach support systems where mothers are treated individually with a "high-risk approach" that helps strengthen their resilience.

CONCLUSIONS

A beneficial way to reduce maternal anxiety in the 12 areas identified would be to develop support systems that provide continuous support for children's mental health care needs, psychological guidance, community support for maternal empowerment, outreach for individual support, and professional consultation for mothers who have high anxiety about radioactivity.

LIMITATIONS

Because this study did not employ a control group, we cannot compare our findings with mothers who reside in non-disaster areas.

ACKNOWLEDGEMENTS

This study was funded by a Health and Labour Sciences Research Grant: 2012-Next Generation-designated-006 (Restoration; Kunihiro Okamura, Principal Investigator) on the theme "Study of health care for pregnant and parturient women in disasters: Public health issues".

REFERENCES

1. Ushizawa, H., Ruth, F.A., Bice, S., Matsui, T., Ueki, Y., Tosaka, N., et al. (2013) Needs for Disaster Medicine: Lessons from the Field of the Great East Japan Earthquake. Western Pacific Surveillance and Response Journal, 4, 51-55.http://dx.doi.org/10.5365/wpsar.2012.3.4.010
2. Umihara, J. and Nishikitani, M. (2013) Emergent Use of Twitter in the 2011 Tohoku Earthquake. Pre-Hospital and Disaster Medicine, 24, 1-7.
3. Omama, S., Yoshida, Y., Ogasawara, K., Ogawa, A., Ishibashi, Y., Nakamura, M., et al. (2013) Influence of the Great East Japan Earthquake and Tsunami 2011 on Occurrence of Cerebrovascular Diseases in Iwate, Japan. Stroke, 44, 1518-1524.http://dx.doi.org/10.1161/STROKEAHA.111.000442
4. Suda, S., Inoue, K., Inoue, K., Sato, K., Saito, H., Matsumoto, T., et al. (2013) Effects of Psychological Distress Due to the Great East Japan Earth Quake, Tsunami, Fukushima Nuclear Power Plant Disasters on Psychiatric Symptoms in Patients with Mental Disorders: Observational Studies in Tochigi. Seishin Shinkeigaku Zasshi, 115, 499-504.
5. Nagata, T., Halamka, J., Himeno, S., Himeno, A., Kennochi, H. and Hashizume, M. (2013) Using a Cloud-Based Electronic Health Record during Disaster Response: A Case Study in Fukushima. Prehospital and Disaster Medicine, 28, 383-387. http://dx.doi.org/10.1017/S1049023X1300037X
6. Galea, S., Brewin, C.R., Gruber, M., Jones, R.T., King, D.W., King, L.A., et al. (2007) Exposure to Hurricane-Related Stressors and Mental Illness after Hurricane Katrina. Archives of General Psychiatry, 64, 1427-1434.http://dx.doi.org/10.1001/archpsyc.64.12.1427

7. Kessler, R.C., Galea, S., Jones, R.T. and Parker, H.A. (2006) Mental Illness and Suicidality after Hurricane Katrina. Bulletin of the World Health Organization, 84, 930-939.http://dx.doi.org/10.2471/BLT.06.033019
8. Sastry, N. and Van Landingham, M. (2009) One Year Later: Mental Illness Prevalence and Disparities among New Orleans Residents Displaced by Hurricane Katrina. American Journal of Public Health, 99, S725-S731. http://dx.doi.org/10.2105/AJPH.2009.174854
9. Jones-DeWeever, A. (2008). Women in the Wake of the Storm: Examining the Post-Katrina Realities of the Women of New Orleans and the Gulf Coast. Institute for Women's Policy Research, Washington DC.
10. Zwiebach, L., Rhodes, J. and Roemer, L. (2010) Resource Loss, Resource Gain, and Mental Health among Survivors of Hurricane Katrina. Journal of Traumatic Stress, 23, 251-258. http://dx.doi.org/10.1002/jts.20579
11. Howitt, D. and Cramer, D. (2010) Introduction to Research Methods in Psychology. 3rd Edition, Prentice Hall, London.
12. Funayama, M. and Mizushima, J. (2013) Severity of Pre-Existing Psychiatric Illness and Response to the Great East Japan Earthquake. Journal of Psychiatric Research, 47, 1479-1482. http://dx.doi.org/10.1016/j.jpsychires.2013.06.006
13. Sato, K. (2013) Influence of the Great East Japan Earthquake on the Mental Health of Mothers. J Tohoku Med Soc, 125, 130-133.
14. Usami, M., Iwadare, Y., Kodaira, M., Watanabe, K., Aoki, M., Katsumi, C., et al. (2012) Relationships between Traumatic Symptoms and Environmental Damage Conditions among Children 8 Months after the 2011 Japan Earthquake and Tsunami. PLoS ONE, 7, e50721. http://dx.doi.org/10.1371/journal.pone.0050721
15. Sugiura, H., Akahane, M., Ohkusa, Y., Okabe, N., Sano, T., Jojima, N., et al. (2013) Prevalence of Insomnia among Residents of Tokyo and Osaka after the Great East Japan Earthquake: A Prospective Study. Interactive Journal of Medical Research, 2, e2.http://dx.doi.org/10.2196/ijmr.2485

16. Takeda, T., Tadakawa, M., Koga, S., Nagase, S. and Yaegashi, N. (2013) Premenstrual Symptoms and Posttraumatic Stress Disorder in Japanese High School Students 9 Months after the Great East-Japan Earthquake. Tohoku Journal of Experimental Medicine, 230, 151-154. http://dx.doi.org/10.1620/tjem.230.151
17. Jones, S.L. and Schmidt, C.K. (2013) Psychosocial Effects of Disaster in Children and Adolescents: Significance and Management. Nursing Clinics of North America, 48, 229-239. http://dx.doi.org/10.1016/j.cnur.2013.01.005
18. Hayashi, M. (2012) Post Major Disaster Mental Health Care Measures: Experiences of Sendai City. Japanese Bulletin of Social Psychiatry, 21, 308-314.
19. Umezaki, S. and Arachi, K. (2012) Implementation and Examination of the "Mental Health Care Team" in the Great East Japan Earthquake. Japanese Journal of Disaster Medicine, 17, 221-224.

Chapter 5

Household and Community Disaster Preparedness in Japanese Provincial City: A Population-Based Household Survey

Jun Tomio[1, 2], Hajime Sato[3], Yuji Matsuda[4], Toshie Koga[5], and Hiroko Mizumura[6]

[1]Department of Disaster Medical Management, University of Tokyo Hospital, University of Tokyo, Tokyo, Japan

[2]Department of Preventive Medicine, St. Marianna University School of Medicine, Kawasaki, Japan

[3]Department of Health Policy and Technology Assessment, National Institute of Public Health, Wako, Japan

[4]Division of Life Sciences, Human Environmental Sciences, Graduate School of Humanities and Sciences, Ochanomizu University, Tokyo, Japan

[5]Department of Architecture and Environmental Design, Kanto Gakuin University, Yokohama, Japan

[6]Department of Human Environment Design, Faculty of Human Life Design, Toyo University, Asaka, Japan

ABSTRACT

Householdand community-level preparedness have been re-emphasized after recent major earthquakes and tsunamis in Japan. The paper examines the prevalence and the determinants of disaster preparedness among the residents of a provincial city in Japan at both levels. Furthermore, it seeks to uncover the associations between householdand community-level preparedness activities to test the hypothesis that a complementary relationship exists between them. We used a subset of a population-based household questionnaire survey of 4000 randomly sampled households in Komoro City in the Nagano Prefecture of Japan in February and March of 2011. The questionnaire included specific questions to measure disaster the preparedness status at both the household and community levels. The characteristics and associations of householdand community-level preparedness were analyzed with multivariable logistic regression models. We found insufficient disaster preparedness at both household and community levels. Older, female, and better educated household heads were associated with better household preparedness, while length at residence, non-single status, presence of an elderly household member, and farming occupations were associated with better community preparedness. Households with one or more household-level preparedness measures were more likely to receive community assistance than those lacking them. The relationship between household and community preparedness was not complementary. Hence, a large proportion of the households were unprepared at both the community and household levels.

INTRODUCTION

Strengthening disaster preparedness is a top priority of disaster management programs at both national and local levels (United

Nations International Strategy for Disaster Reduction, 2007). However, recent disasters have revealed limitations in the timing and mobility of government assistance to the public. Therefore, preparedness plans for householdsthat are based on the individual and/or household responsibility and those for communitiesthat are based on the mutual assistances within the community, have been highlighted as areas for improvement (Central Disaster Management Council, 2011). The emergency responses of local and national governments often cannot reach affected populations immediately after an event, especially when a disaster strikes a large area at the same time. Hence, the preparedness of individuals, households, and communities, along with that of the government, is crucial for improving community resilience in the face of a disaster (Central Disaster Management Council, 2011; Diekman, Kearney, O'Neil, & Mack, 2007; Kapucu, 2008; Norris, Stevens, Pfefferbaum, Wyche, & Pfefferbaum, 2008). Several events associated with the Great East Japan Earthquake (March 2011) have confirmed the importance of community preparedness (e.g., mutual aid in neighborhoods, schools, and communities) (Japanese Red Cross Society, 2011; Matanle, 2011; Nakahara, 2011).

Previous studies (Bethel, Foreman, & Burke, 2011; Eisenman et al., 2006; Kapucu, 2008; Tomio, Sato, & Mizumura, 2010, 2011; Uscher-Pines et al., 2009) have assessed various determinants of household preparedness, including the level of risk awareness (Eisenman et al., 2006; Murphy, Cody, Frank, Glik, & Ang, 2009) and the age of household heads (Baker, 2011; Eisenman et al., 2006; Murphy et al., 2009), indicating that preparedness activities at the household level are typically not undertaken, although some studies provide mixed results. Individual citizens should plan for real disasters in order to help not only their families but also their friends and neighbors when a catastrophe or an emergency strikes (Kapucu, 2008). However, previous findings have revealed that many households are not ready for a disaster. For example, adequate supplies of food and water are not present in them. To ensure that individual households achieve an adequate level of preparedness, those who are poorly organized must rely on community-level preparedness. For example, they must be able to identify someone in the neighborhood who can provide assistance in the event of a disaster. Accordingly, there should be a complementary relationship between householdand community-level preparedness, i.e., a lower level of household preparedness would be associated with a higher level of community preparedness, and vice versa. Some studies indicate common predictors of different levels of preparedness, such as people's subjective norms(Motoyoshi, Takao, & Ikeda, 2008; Paek, Hilyard, Freimuth, Barge, & Mindlin, 2010), and concern about disasters (Eisenman et al., 2006; Motoyoshi et al., 2008), and connectedness to a story-telling network (Kim & Kang, 2010), indicating a positive rather than a complementary relationship between the two types of preparedness. However, the association has not been fully understood. Identifying the pattern of the association between householdand community-level preparedness should provide relevant information for community and local authorities when they decide to approach local residents.

The objectives of the present study are to examine the prevalence and the determinants of disaster preparedness among the residents of a provincial city in Japan at both household and community levels and to analyze the associations between these two levels of preparedness to test the hypothesis that they have a complementary relationship.

MATERIALS AND METHODS

Survey and Study Setting

In the present study, we used a subset of a population-based household survey conducted in Komoro City in Nagano Prefecture of Japan in February 2011. The survey was designed to describe citizen attitudes toward injury and disaster prevention and citizen consciousness of safety and security as part of the city's new safety promotion policy. Komoro is a small-sized city in central Honshu, the main island of Japan, with a population of approximately 44,000 as of September 2011. Its total area is 98.66 km^2, and the city is situated at the foot of Mount Asama, an active complex volcano. Some rivers run through the city, and the major natural hazards of the area include earthquakes, volcanic eruptions, and floods. Fortunately, a lengthy period has transpired since a disastrous event affected the area (Komoro Municipal Disaster Management Council, 2010).

The survey's target population comprised 17,300 households, covering all but one street in Komoro City. Of the total number of households, 4000 were sampled randomly and proportionately by street, using the Basic Resident Registration System, Japan's official means of recording residents, as of September 1, 2010. The survey questionnaires were mailed to the 4000 randomly selected households in February 2011. The head of each household was asked to complete the questionnaire (anonymously) and return it to the Komoro City Hall by the end of February. A separate survey was conducted with all 406 households in Yachihara, the area designated as the target for the all-household survey and exempted from the original survey. The same questionnaire and methods were used, but the second survey was administered in March 2011, just after the Great East Japan Earthquake of March 11 and the North Nagano Prefecture Earthquake of March 12 (magnitude of 6.7 on the Richter scale, with an epicenter 70 km north of Komoro City). The local government of Komoro City and the Ethical Committee of St. Marianna University School of Medicine and Toyo University reviewed and approved the protocol for conducting the study and the questionnaire.

Study Variables

The questionnaire included two specific questions originally developed to measure disaster preparedness at both household and community levels. Respondents chose one or more appropriate responses from the following basic preparedness activities to describe the household preparedness level: 1) securing furniture; 2) stockpiling food and water; 3) locating the designated evacuation center; 4) preparing an emergency kit; and 5) discussing disaster responses with family members. We established that a household had also adopted procedures reflecting community-level preparedness by determining if its head answered "yes" to the following question: "Do you have any neighborhood inhabitants who would help each other in times of disaster?"

The survey included questions regarding the basic characteristics of households, including area of residence (central or suburb), presence or absence of an officially accredited voluntary disaster preparedness organization (VDPO) that engages the community in disaster drills and other preparedness activities, length of residence (<5 years, 5 - 19 years, or 20+ years), and household size (single or not). Further, respondents were asked to indicate the presence of household member(s) under 15 and over 74 years of age, those requiring special assistance due to disabilities, and those receiving long-term care. Households were identified as concerned about natural disasters if they indicated "natural disaster" as one of the top 3 of 15 items that would be relevant for their members in the next five years in terms of safety promotion, such as traffic accidents or criminal activities. Some relevant characteristics of the household heads, including gender, age (under 50, 50 - 64, or 65 years or older), educational level (college graduate or not), and occupational status (employed/self-employed, farmer, or unemployed/retired/housewife), were also collected.

Statistical Analysis

To describe the status of preparedness, we calculated the proportion of households that had adopted household and community levels preparedness measures. Five activities of household-level preparedness were summarized and categorized as follows: none, one, or two or more. They were categorized dichotomously, however, when treating preparedness at the household level as a dependent variable in the logistic regression model. Associations among the household characteristics described above and both householdand community-level preparedness were analyzed with multivariable logistic regression models to determine possible predictors of each level of preparedness. The association between householdand community-level preparedness was analyzed with the chi-squared test, followed by the application of a multivariable logistic regression model that controlled for household-related characteristics. Additionally, we analyzed the data from the Yachihara area, collected after the major events of March 2011, using the above-mentioned methods to identify any major changes in preparedness status immediately after the event. All statistical analyses were performed using Stata version 11.2 (StataCorp, College Station, TX).

RESULTS

Overall, 1985 (50%) valid responses regarding preparedness status were obtained from the targeted 4000 households. Of these, 41% lived in the center of the city, 57% in a street with VDPO, 15% in single households, 24% with children under 15 years, 19% with elderly aged

75 years or older, 17% with one or more family members requiring special assistance; 75% had lived in the city for 20 years or longer, and 26% indicated concern about natural disaster (Table 1). A majority of the heads of households were men, aged 50 years or older, with education levels below that of college graduate; they were generally employees or self-employed.

Table 1: Disaster preparedness and characteristics of the study subjects (n = 1985)

Variable	n (%)
Household level preparedness	
Measures to secure furniture	365 (18)
Food and water stockpiling	542 (27)
Identifying the designated evacuation center	651 (33)
Preparing emergency pack	224 (11)
Discussing disaster response with family	197 (10)
Potential mutual aid in the neighborhood	1105 (56)
Area of residence*	
Center	777 (41)
Suburb	1127 (59)
Voluntary disaster preparedness organization*	1002 (57)
Years of residence*, year	
<5	147 (7)
5 - 20	341 (17)
20+	1481 (75)
Single household*	288 (15)
Living with children under age 15	486 (24)
Living with someone aged 75 years or older	369 (19)
Living with those requiring special assistance	345 (17)

Male*	1401 (71)
Age*, year	
<50	611 (31)
50 - 64	786 (40)
65+	583 (29)
College graduate*	690 (35)
Occupational status*	
Employee/self-employed	1249 (63)
Farming	187 (9)
Unemployed/retired/housewife	538 (27)
Concerned about natural disasters	508 (26)

*Variables include some missing data

Concerning household-level preparedness, 18% had secured their furniture and 27% had stockpiled food and water; 33% correctly identified the designated evacuation center; 11% had prepared emergency kits; and 10% had discussed disaster response with family members (Table 1). More than half of the households (56%) reported that they could rely on their neighbor(s) for potential mutual aid in case of a disaster.

Households were more likely to be prepared at the household level when the heads of household were women (Odds ratio [OR] 1.40, 95% confidence interval [CI] 1.11 - 1.77), older (OR 1.43, 95% CI 1.03 - 1.97 for 50 - 64 years and OR 1.97, 95% CI 1.35 - 2.87 for 65 years or older), or college graduates (OR 1.55, 95% CI 1.22 - 1.96) than their counterparts (Table 2). However, those with longer years of residence (OR 1.65, 95% CI 1.03 - 2.62 for 5 - 19 years, and OR 3.08, 95% CI 1.99 - 4.76 for 20+ years) and who lived with persons aged 75 years or older (OR 1.40, 95% CI 1.06 - 1.83) were more likely to have neighbors on whom they could rely for potential mutual aid than their counterparts. Single households, however, were less likely to rely on neighbors for potential aid (OR 0.59, 95% CI 0.44 - 0.80). Regarding the occupational status of the heads of households, farmers were more likely to have someone they could rely on for potential mutual aid than employees and the self-employed (OR 2.64, 95% CI 1.70 - 4.10).

The proportion of households with access to potential mutual aid during a disaster increased according to the household preparedness level. More than two-thirds of the households that had described two or more activities relevant to household-level preparedness were aware of possible community resources, while less than half of those that had not engaged in such activities ($P < 0.001$) felt they had access to community resources for aid (Table 3). The associations were also found after controlling for the effects of household-related factors and the socio-demographic status of the household heads. Households with either one item and/or two or more items associated with household-level preparedness were more likely to receive community assistance than those with no evidence of household preparedness (OR 1.40, 95% CI 1.11 - 1.77, and OR 2.31, 95% CI 1.78 - 3.00, respectively).

Table 2: Associations between household characteristics and disaster preparedness at the household level and potential mutual aid (n = 1728)

Variable	**Household preparedness**	**Potential mutual aid**
	Adjusted OR (95% CI)	Adjusted OR (95% CI)
Living in suburb	0.86 (0.69 - 1.08)	1.22 (0.99 - 1.51)
VDPO	1.05 (0.84 - 1.31)	1.13 (0.93 - 1.39)
Length of residence, years		
<5	Reference	Reference
5 - 19	1.15 (0.68 - 1.93)	1.65 (1.03 - 2.62)*
20+	1.07 (0.66 - 1.74)	3.08 (1.99 - 4.76)**
Single household	0.87 (0.62 - 1.21)	0.59 (0.44 - 0.80)**

Living with children under age 15	0.95 (0.69 - 1.29)	1.06 (0.80 - 1.39)
Living with someone aged 75 or older	1.21 (0.91 - 1.61)	1.40 (1.06 - 1.83)*
Living with those requiring special assistance	0.92 (0.69 - 1.24)	0.86 (0.66 - 1.13)
Female	1.28 (1.00 - 1.64)*	0.95 (0.76 - 1.19)
Age, year		
<50	Reference	Reference
50 - 64 yr	1.43 (1.03 - 1.97)*	1.00 (0.75 - 1.33)
65 yr+	1.97 (1.35 - 2.87)**	1.31 (0.93 - 1.85)
College graduate	1.55 (1.22 - 1.96)**	1.22 (0.93 - 1.52)
Occupational status		
Employee/self-employed	Reference	Reference
Farming	1.01 (0.67 - 1.51)	2.64 (1.70 - 4.10)**
Unemployed/ retired/ housewife	1.01 (0.76 - 1.35)	0.79 (0.61 - 1.04)
Concerned about natural disaster	1.05 (0.83 - 1.34)	1.18 (0.94 - 1.48

*$P < 0.05$, **$P < 0.01$. Household preparedness was regarded as positive when the households adopted two or more activities. The reference value of the odds ratio for each bivariate variable is the counterpart value. VDPO: voluntary disaster preparedness organization; OR: odds ratio; CI: confidence interval.

Table 3: Association between household preparedness status and potential mutual aid

Number of adopted household preparedness activities		**Households with potential mutual aid**	**Adjusted OR (95 % CI)**
Number of activities	n	n (%)	
None	770	365 (47)	Reference
1	703	390 (55)	1.40 (1.11 - 1.77)*
2+	512	350 (68)	2.31 (1.78 - 3.00)*

*P < 0.001 with multivariable logistic regression model adjusted for the characteristics of the households and household heads shown in Table 2.

A greater proportion of households in Yachihara (where the data was collected immediately following the catastrophe of March 11, 2011) were prepared for disasters, that is, they had adopted two or more activities associated with household-level preparedness, compared to the rest of the city (43% and 26%, respectively). However, no difference between the residents of Yachihara and the rest of Komoro City was found in terms of access to mutual aid (56% for both). Similar to the findings on the households in the rest of the city, our results indicated that households characterized by either two or more activities or one preparedness activity were more likely to receive mutual aid than those with no household-level preparedness activity (OR 5.09, 95% CI 1.71 - 15.12, and OR 3.27, 95% CI 1.08 - 9.86, respectively).

DISCUSSION

The results of the population-based survey administered in a provincial city in Japan revealed insufficient disaster preparedness at both household and community levels. Older age, female, and better-educated household heads were associated with better household preparedness, while length of residence (at least five years), nonsingle status, the presence of an elderly household member, and farming occupations were associated with better community preparedness. The relationship between household and community preparedness was not complementary; hence, a large proportion of the households were unprepared at both the community and household levels.

Household disaster preparedness has been assessed in various settings by different measures. Previous studies and surveys have reported a low occurrence of preventive measures in communities. Our findings were similar to the Special Opinion Poll on Disaster Preparedness for Earthquake conducted by the Cabinet Office of Japan in 2007 on the country's general population, which indicated that 36% of respondents had stockpiled food and water, 24% had stabilized furniture, and 23% were prepared to carry valuables and other supplies with them if forced to evacuate (Cabinet Office, 2007). Several recent studies conducted in Japan (Mimaki, Takeuchi, & Shaw, 2009), the US (Eisenman et al., 2006), and Italy (Miceli, Sotgiu, & Settanni, 2008) have reported similar levels of preparedness activities in these countries. Other studies have, however, showed better preparedness levels. For example, an American study that utilized the Behavioral Risk Factor Surveillance System (BRFSS) indicated that over 80% of the households has stockpiled food (Bethel et al., 2011). A study conducted in a hurricane-prone area in Florida (Baker, 2011)arrived at similar results. The preparedness of the present study's Japanese subjects can be characterized as within the normal range; however, the fact that less than a third of the households adopted basic preparedness activities should be regarded as insufficient, considering the high risk of disasters in Japan.

The results of the multivariable logistic regression analysis suggested that older age, female, and better-educated household heads are potential predictors of better household preparedness. These results support several previous findings (Baker, 2011; Eisenman et al., 2006; Levac, Toal-Sullivan, & O Sullivan, 2011; Murphy et al., 2009); however, they may be confounded by other factors, especially financial status, which has been indicated as a strong predictor of material preparedness at the household level (Baker, 2011; Levac et al., 2011; Murphy et al., 2009). Financial status is generally higher when the household head is older in Japan (Ministry of Internal Affairs and Communications, 2010).

Potential resources for mutual aid were found in more than half of the households, though we cannot make a relative evaluation of this result because previous studies did not investigate the status of mutual aid at the community level. The results of multivariable logistic regression analysis indicated that households in which residents had lived for at least five years and those with two or more family members, elderly family members, and farmers as heads of households were likely to receive mutual aid in the event of a disaster. In fact, these variables were considered as predictors of a "neighborhood bond", according to the results of the National Survey on Lifestyle Preferences 2007 conducted by the Cabinet Office of Japan (Quality-of-Life Policy Bureau (Cabinet Office), 2007). "Neighborhood belonging", similar in concept to the neighborhood bond, was found to be a possible predictor of the actual preparedness actions of the community affected by Hurricane Ivan in 2004 (Kim & Kang, 2010). Thus, the strength of the neighborhood bond may be regarded as a forecaster of potential mutual aid in a community based on the household characteristics listed above. At the same time, the local government should pay special attention to single residents and/or those who live in the community for short periods, as they could be at high risk for slipping through the network of community preparedness.

The existence of a VDPO in the community was not associated with the higher preparedness of households and communities, and this finding is contrary to a previous report on a local coastal city in Japan (Mimaki et al., 2009). Although a VDPO may have beneficial effects that were not measured in the present inquiry, some studies argue that certain VDPOs are not rooted in the community and have become ineffective (Takahashi, 1995). The role of VDPOs in disaster preparedness should therefore be re-examined.

Interestingly, respondents in our study who expressed concern about natural disasters did not prepare more for a major event than those who expressed less concern. Previous studies have indicated that higher risk awareness and higher risk perception predicted better household preparedness (Eisenman et al., 2006; Kapucu, 2008; Murphy et al., 2009). Although the concern about disasters measured in the present study could be partially translated into risk awareness, it may be merely a vague feeling of anxiety that would not lead to preparedness behavior. Risk awareness and perception translate into preparedness action when they are based on individual understandings of the hazard (Kapucu, 2008). Thus, further research should investigate hazards and vulnerability to understand the reasons for individual concerns and the role of disaster risk communications.

Contrary to our hypothesis on the existence of a complementary relationship between household and community preparedness, we observed that households exhibiting readiness were more likely to have access to potential resources for mutual aid in times of disaster. The supplemental analysis revealed the same pattern of association among residents in the Yachihara area.

Intuitively, people promote community and/or public emergency preparedness in order to seek help from neighbors and local authorities when they cannot take appropriate prepare at the household level. The results of the present study showed a positive interaction between preventive measures at the household and community levels rather than a complementary relationship. Murphy et al. have reported a similar positive interaction between individual and public preparedness, indicating that increased confidence in the ability of local health care systems to handle emergencies appeared to correlate with the proactive activities of individuals (Murphy et al., 2009).

Some possible explanations for this positive relation can be proposed. People living in well-prepared households may build networks of mutual aid among neighbors. Moreover, people living in a neighborhood with existing mutual aid may more easily obtain information about household preparedness. In addition, other common root causes may exist at both levels of preparedness, although the present study did not find a common predictor. Earlier research has found that a person's subjective norms or his or her beliefs of the sort of behaviors that significant others (e.g., family members and close friends) expect determined preparedness (Motoyoshi, Takao, & Ikeda, 2004; Motoyoshi et al., 2008; Paek et al., 2010). Motoyoshi et al. have observed the influence of subjective norms on the attitudes of Japanese residents toward household and community preparedness, although they did not examine the association between the two levels (Motoyoshi et al., 2008). This finding implies the existence of discrepancy in the subjective norms for disaster preparedness among local residents. Further studies need to examine this assumption.

Supplemental analyses of the Yachiahra area have shown an increased level of household preparedness after major earthquakes; however, no significant change was observed for the households with potential mutual aid. Although the present study could not measure the effect of major events, this observation suggests that the experiential effect of a major event may differ according to the level of preparedness. Thus, separate approaches might be required to improve the different types of preparedness, even though they share some common root causes.

This study has several limitations. First, a high response rate was not achieved, although the community survey had a 50% response rate from a sample group that was demographically similar to the target population. Second, because of the nature of the cross-sectional data, the findings reflect associations rather than causation. Third, preparedness activities included in the survey were not comprehensive, and they were not analyzed qualitatively. Community preparedness, for example, was measured only by the existence of potential mutual aid; other measures, such as participation in disaster drills, were not examined. Fourth, the study did not control for some known predictors of preparedness, including financial status and disaster risk perception. Lastly, the survey was conducted in an inland provincial city in Japan; therefore, the results might not be applicable to locations with different geographical characteristics, such as large cities, coastal areas, and areas that are highly or minimally prone to disasters.

CONCLUSIONS

In summary, we found that community-level preparedness did not appear to complement preparedness at the household level. Instead, the findings suggested the polarization of disaster preparedness in the community. Policy makers should be aware of high-risk subgroups that are poorly prepared for disasters at both household and community levels. Although we did not find common determinants of poor preparedness at both levels, we suggest that focusing on easily detectable high-risk individuals, e.g., those who have recently started to live in a city and those who are single, could be a good starting point for intervention. Local authorities should be encouraged to develop educational programs that promote preparedness behaviors based on sound disaster risk awareness. In addition, each community should be required to identify publicly key community members, such as social workers, who could provide support to neighbors in the event of a disaster. We highly recommend that further research examines the determinants of poor preparedness and the quality of preventive activities and identify effective approaches for improving preparedness at all levels.

ACKNOWLEDGEMENTS

We would like to thank the Komoro citizens and Komoro city officials who are in charge of the Komoro Safe Community. This study was supported in part by the Grant-in-Aid for Young Scientists (B) (24790497), the Grant-in-Aid for Challenging Exploratory Research (25560395), the Grant-in-Aid for Scientific Research (C) (24590827) and the Grant-in-Aid for Scientific Research (B) (22300254) from the Japan Society for the Promotion of Science.

REFERENCES

1. Baker, E. J. (2011). Household Preparedness for the Aftermath of Hurricanes in Florida. Applied Geography, 31, 46-52. http://dx.doi.org/10.1016/j.apgeog.2010.05.002
2. Bethel, J. W., Foreman, A. N., & Burke, S. C. (2011). Disaster Preparedness among Medically Vulnerable Populations. American Journal of Preventive Medicine, 40, 139-143.http://dx.doi.org/10.1016/j.amepre.2010.10.020
3. Cabinet Office (2007). Special Opinion Poll for Disaster Preparedness for Earthquake. Cabinet Office.

4. Central Disaster Management Council (2011). Basic Disaster Management Plan.http://www.bousai.go.jp/keikaku/20111227_basic_plan.pdf
5. Diekman, S. T., Kearney, S. P., O'Neil, M. E., & Mack, K. A. (2007). Qualitative Study of Homeowners' Emergency Preparedness: Experiences, Perceptions, and Practices. Prehospital and Disaster Medicine, 22, 494-501.
6. Eisenman, D. P., Wold, C., Fielding, J., Long, A., Setodji, C., Hickey, S., & Gelberg, L. (2006). Differences in Individual- Level Terrorism Preparedness in Los Angeles County. American Journal of Preventive Medicine, 30, 1-6.http://dx.doi.org/10.1016/j.amepre.2005.09.001
7. Japanese Red Cross Society (2011). News & Stories: Rikuzentakata Fights for Survival. Japanese Red Cross Society.
8. Kapucu, N. (2008). Culture of Preparedness: Household Disaster Preparedness. Disaster Prevention and Management, 17, 526-535.http://dx.doi.org/10.1108/09653560810901773
9. Kim, Y.-C., & Kang, J. (2010). Communication, Neighbourhood Belonging and Household Hurricane Preparedness. Disasters, 34, 470-488. http://dx.doi.org/10.1111/j.1467-7717.2009.01138.x
10. Komoro Municipal Disaster Management Council (2010). Komoro Local Disaster Management Plans. http://www.city.komoro.nagano.jp/bosai/index.htm
11. Levac, J., Toal-Sullivan, D., & O'Sullivan, T. L. (2011). Household Emergency Preparedness: A Literature Review. Journal of Community Health, 37, 725-733.http://dx.doi.org/10.1007/s10900-011-9488-x
12. Matanle, P. C. D. (2011). The Great East Japan Earthquake, Tsunami, and Nuclear Meltdown: Towards the (Re) Construction of a Safe, Sustainable, and Compassionate Society in Japan's Shrinking Regions. Local Environment, 16, 823-847.http://dx.doi.org/10.1080/13549839.2011.607160
13. Miceli, R., Sotgiu, I., & Settanni, M. (2008). Disaster Preparedness and Perception of Flood Risk: A Study in an Alpine Valley in Italy. Journal of Environmental Psychology, 28, 164-173. http://dx.doi.org/10.1016/j.jenvp.2007.10.006

14. Mimaki, J., Takeuchi, Y., & Shaw, R. (2009). The Role of Community-Based Organization in the Promotion of Disaster Preparedness at the Community Level: A Case Study of a Coastal Town in the Kochi Prefecture of the Shikoku Region, Japan. Journal of Coastal Conservation, 13, 207-215. http://dx.doi.org/10.1007/s11852-009-0065-8
15. Ministry of Internal Affairs and Communications (2010). Family Income and Expenditure Survey. Ministry of Internal Affairs and Communications.
16. Motoyoshi, T., Takao, K., & Ikeda, S. (2004). Determinant Factors of Community-Based Disaster Preparedness: A Case Study of Flood Prone Area. Japanese Journal of Psychology, 75, 72-77. http://dx.doi.org/10.4992/jjpsy.75.72
17. Motoyoshi, T., Takao, K., & Ikeda, S. (2008). Determinants of Householdand Community-Based Disaster Preparedness. The Japanese Journal of Social Psychology, 23, 209-220.
18. Murphy, S. T., Cody, M., Frank, L. B., Glik, D., & Ang, A. (2009). Predictors of Emergeny Preparedness and Compliance. Disaster Medicine and Public Health Preparedness, 3, 1-10.
19. Nakahara, S. (2011). Lessons Learnt from the Recent Tsunami in Japan: Necessity of Epidemiological Evidence to Strengthen Community-Based Preparation and Emergency Response Plans. Injury Prevention, 17, 361-364. http://dx.doi.org/10.1136/injuryprev-2011-040163
20. Norris, F. H., Stevens, S. P., Pfefferbaum, B., Wyche, K. F., & Pfefferbaum, R. L. (2008). Community Resilience as a Metaphor, Theory, Set of Capacities, and Strategy for Disaster Readiness. American Journal of Community Psychology, 41, 127-150.http://dx.doi.org/10.1007/s10464-007-9156-6
21. Paek, H. J., Hilyard, K., Freimuth, V., Barge, J. K., & Mindlin, M. (2010). Theory-Based Approaches to Understanding Public Emergency Preparedness: Implications for Effective Health and Risk Communication. Journal of Health Communication, 15, 428-444.http://dx.doi.org/10.1080/10810731003753083
22. Quality-of-Life Policy Bureau (Cabinet Office) (2007). National Survey on Lifestyle Preferences FY2006. Quality-of-Life Policy Bureau (Cabinet Office).

23. Takahashi, K. (1995). A Study on the Present Situations and Problems of Voluntary Organizations for Disaster Prevention in Nagasaki City after the 1982 Nagasaki Flood Disaster. Journal of Japan Society for Natural Disaster Science, 14, 219-234.

24. Tomio, J., Sato, H., & Mizumura, H. (2010). Interruption of Medication among Outpatients with Chronic Conditions after a Flood. Prehospital and Disaster Medicine, 25, 42-50.

25. Tomio, J., Sato, H., & Mizumura, H. (2011). Disparity in Disaster Preparedness among Rheumatoid Arthritis Patients with Various General Health, Functional, and Disability Conditions. Environmental Health and Preventive Medicine, 17, 322-331. http://dx.doi.org/10.1007/s12199-011-0257-3

26. United Nations International Strategy for Disaster Reduction (2007). Hyogo Framework for Action 2005-2015: Building the Resilience of Nations and Communities to Disasters. United Nations International Strategy for Disaster Reduction.

27. Uscher-Pines, L., Hausman, A. J., Powell, S., DeMara, P., Heake, G., & Hagen, M. G. (2009). Disaster Preparedness of Households with Special Needs in Southeastern Pennsylvania. American Journal of Preventive Medicine, 37, 227-230.http://dx.doi.org/10.1016/j.amepre.2009.04.028

Chapter 6

Factors Influencing Female Patients' Recovery after their First Myocardial Infarction as Experienced by Cardiac Rehabilitation Nurses

Inger Wieslander[1, 2], Jan Mårtensson[2], Bengt Fridlund[2], and Petra Svedberg[1]

[1]School of Social and Health Sciences, Halmstad University, Halmstad, Sweden

[2]School of Health Sciences, Jönköping University, Jönköping, Sweden

ABSTRACT

Background

In the developed part of the world, coronary heart disease is the major cause of death and is one of the leading causes of disease burden. In Sweden, more than 30,000 people per year are affected by myocardial infarction and out of these approximately 40% are women. Nearly 70% of the women survive and after a myocardial infarction a recovery process follows. Today's health care focuses more on treatment, symptoms and risk factors than on the individuals' perceptions of the recovery process.

Aim

To explore cardiac rehabilitation nurses' experiences of factors influencing female patients' recovery after their first myocardial infarction.

Methods

Twenty cardiac rehabilitation nurses were interviewed. The study was conducted using qualitative content analysis.

Results

The cardiac rehabilitation nurses experienced that women's recovery after a first myocardial infarction was influenced whether they had a supportive context, their ability to cope with the stresses of life, if they wanted to be involved in their own personal care and how they related to themselves.

Conclusions

Women's recovery after a myocardial infarction was influenced by factors related to surroundings as well as own individual factors.

The underlying meaning of women's recovery can be described as the transition process of a recovery to health. Our findings suggest that a focus on personcentered nursing would be beneficial in order to promote the every woman's personal and unique recovery after a myocardial infarction. Finally, the cardiac rehabilitation nurses' experiences of factors in- fluencing male patients' recovery after their first myocardial infarction should be important to investigate.

INTRODUCTION

In the developed part of the world, coronary heart disease (CHD) is the major cause of death and is one of the leading causes of disease burden [1]. Myocardial infarction (MI) has long been seen as a male disease despite the fact that it is also a health problem for women [2]. In Sweden, more than 30,000 people per year are affected by MI and out of these approximately 40% are women. Nearly 70% of the women survive [3] and after an MI a recovery process follows [4].

European guidelines on cardiovascular health care emphasize the inclusion of secondary prevention [5]. It is important to note that secondary prevention care after an MI has not improved during the last 15 years at the same rate as acute MI care [6]. One reason could be that research and health care focus more on treatment, symptoms and risk factors and not on the individuals' perceptions of the recovery process after an MI [4]. Today's health care requires both preventative and health promotion interventions which emphasize that it is no longer enough to focus only on curing and alleviating disease [7]. A central part of secondary prevention is cardiac rehabilitation programs [CRP] in order to improve quality of life among patients [8]. The goals of CRP are to reduce the risk of relapse, promote lifestyle and hea th behaviour changes as well as to facilitate the patients' recovery [3,5]. Research has shown that participation in CRP promotes patients' health and quality of life as well as preventing new cardiac events [9]. However, women do not participate in CRP to the same extent as men [10, 11]. Women's life situation after an MI is more likely to be affected negatively as they are, to a greater extent, more concerned about new cardiac events [12].

The concept of recovery can be defined in different ways depending on in which context, profession and culture it is used [13]. Recovery and rehabilitation are often used interchangeably [3], even though there is a distinct difference between them. Slade and Hayward [14] describe that recovery includes both personal recovery and clinical recovery. Clinical recovery, as well as the rehabilitation, focuses upon the reduction of symptoms and increased levels of functioning, which are more in line with the professional perspective [14,15]. Personal recovery can be seen as a multidimensional process [15] and as each person's unique journey [14,16,17]. This unique journey is about the individual's ability to adapt to the new situation after a disease and at the same time to have the power to develop their own new life situation. The person needs to have the opportunity to choose different ways in the recovery process in order to gain health [14,16,17]. Research on recovery and MI, shows that patients and health care professionals describe recovery as an on-going and enduring process that includes learning to adapt to a new life situation [18]. Women, despite illness, describe recovery after an MI as a gradual and personal process in which they face challenges with a combination of strengths, weaknesses and available resources to gain health [19].

A cardiac rehabilitation nurse [CRN] is an expert in both health promotion and in illness prevention [20]. In addition to an adequate knowledge of cardiology and cardiovascular nursing, a CRN has a central role in CRP [20, 21]. However, empirical studies concerning nurses' experiences of women's recovery after a first MI are rare. Most previous research focuses on recovery from the women's and their partners' point of view [12, 22-30]. Since CRNs meet many women who have experienced an MI, their experiences of the women's recovery process may be an important complement in the daily work to support counseling female patients through their recovery process after their first MI. Accordingly the aim of the study was to explore cardiac rehabilitation nurses' experiences of the factors that influence female patients' recovery after their first MI.

METHODS

Design and Setting

The study had an explorative and descriptive design based on qualitative content analysis [31]. The interviews were carried out between April and December 2010 at 10 different hospitals (4 university, 3 county and 3 district hospitals) in Sweden. The heads of the cardiology departments were contacted and asked to give permission for cardiac rehabilitation nurses to participate in the study. All 10 Heads gave their approval and forwarded the names of the CRNs who met the inclusion criteria.

Participants

The inclusion criteria were CRNs, who were actively working with comprehensive CRPs in which physical training, education, risk factor modification, medical checks-ups, pharmacological treatment, psychological support and counseling, as well as vocational rehabilitation were included. The CRNs also had to be members of a multidisciplinary cardiac rehabilitation team (e.g. cardiologist, nurse, and physiotherapist) and fully understand, read and speak Swedish. Twenty-one CRNs were informed and asked to participate by the main author (IW), only one declined; due to time constraints. All CRNs were females and ages ranged between 28 and 65. The time having worked in cardiac care was between 4 and 25 years and between 2 and 23 years within cardiac rehabilitation.

Data Collection

Data was collected through interviews conducted and recorded by the main author (IW). The main author phoned the CRNs and asked them to choose a place and time for the interview. All interviews were conducted at the informants' work place. Four pilot interviews were conducted, which then were included in the analysis. The interviews started with an open main question: "What do you think about women's recovery after a first myocardial infarction?" In order to reach depth in data follow up questions such as "What do you think when you say...?" and "Can you explain more about...?" were asked. The interviews lasted between 35 and 75 minutes. All the interviews were transcribed verbatim.

Data Analysis

Data was processed by the main author, in collaboration with the co-authors, using qualitative content analysis which consists of several analysis steps as described by Graneheim and Lundman [32]. The interviews were read through several times to become familiar with the text. Meaning units that were related to the aim of the study were placed in an analytical matrix. These meaning units were then condensed without losing the core of the content. In the next analysis step, the condensed meaning units were abbreviated to codes. These codes were compared, based on differences and similarities, and 12 subcategories were created forming 4 categories at a manifest level. Based on the content of these categories an overall theme emerged, which expressed the latent meaning of the content (Figure 1) [32].

Ethical Consideration

The ethical principles of the study were according to the requirements of the world medical association declaration of Helsinki [33] and the heads of the cardiology departments have approved the study. After the participants were fully informed, both verbally and in writing, of the aim of the study and the voluntary nature of participation, as well as the possibility of withdrawing, an informed consent was obtained.

RESULTS

Being in a Supportive Context

This category described that CRNs' experiences that women's opportunities to receiving support could either help or hinder the women's recovery process. A major requirement for recovery was access to a good social network, where family, relatives, peers and friends had a supportive attitude, as well as support from health care and workplace.

The newly found situation involved not only the woman, but the whole family so therefore it was important how the support from the family was. The women, who had a functioning family life, where they felt safe and harmonious, generally recovered more easily. "The family, home, and how they have it at home, has an impact" ... (n3). It was important that the female's partner was actively involved in the recovery process, as well as a well-functioning dialogue between family members. Interaction with children and grandchildren could distract the women from having negative thoughts. In cases where the children helped with practical things also created better possibilities for the women to recover. Likewise, when the family took into account, that it is not always the woman, the mother, who has to do everything in the family, this influenced the woman's recovery in a positive way. However, the women's recovery was negatively influenced when they lacked support from their families as well as unsympathetic partners and family members who, neither understood, nor took into account the new situation. Many women, I've heard this many times by women who say that "Yes, but you were in hospital for four days, now you're all right. What is wrong with you"? (n1). The families did not take time to listen, but instead, perceived the women as somewhat whining. Furthermore, they wanted everything to work in the home as before and desired the traditional female role. The women had more difficulty to implement the necessary recommended lifestyle changes. However, women became passive, especially in cases where women were overprotected by their family and were not allowed to do any daily work or exercise at all; which did not help the recovery process.

Figure 1: Illustration of the transition process to the recovery to health through theme, categories and subcategories.

Support from work where the women were given possibilities to participate in various rehabilitation groups during working hours affected their recovery positively. "…, it all depends on... what they are working with. Large companies often let them go, but small, private companies, where they often do not have many employees, therefore more dependent upon staff, where it is a little harder to take one afternoon off a week ..." (n3). The women's recovery was facilitated when employers arranged more regular working hours and less planned overtime.

Regarding support from people with the same experience, it emerged that women usually experienced great gains when they met others in the same situation in the rehabilitation groups. "I've heard women say that really, to meet other women in the same situation is better than the entire health care! So it is of great importance" (n2). Friendships developed, for example at the exercise training during the months the group members met, which led the women to continue to meet and exercise together, share experiences and support each other, after the initial group training period had come to an end. The women emphasized the importance of meeting people who understood their situation which was perceived as an additional support, something that made them realize that they were not alone.

Concerning support from female friends, it became evident that women willingly spoke more to their friends than to their partners when it came to their innermost thoughts and their situation. "And then women have more than men, good friends who support each other, I am sure we can say ..." (n13). Since emotional support was important for the women with regard to recovery, it was important that there were female friends around when needed.

Support from healthcare professionals deals with counseling and support during the women's recovery process. This involved: having an experienced nurse to turn to, having confidence and assurance in the nurse, as well as, continuity of care. The CRNs, as well as the team's approach, were also important in regard to the women's recovery. "The fact that we also provide support, that they can call us and get adequate and accurate answers to their, all the questions they have... that we have the expertise, that we have faith to give a correct answer. Both in medical terms, and to be able to encourage and help them when attempting to quit smoking..." (n11). The women's recovery was also influenced when the nurses managed to get the women: motivated to exercise, to be aware of both their risk and health factors which resulted in the women feeling better and stronger after visiting the nurse. Even physical training, under supervision, gave the women security which stimulated them to do more physical exercise. Likewise, medical assistance was of great importance. Their recovery was also affected by them knowing that they were prescribed the appropriate medication, and that the physician spoke about what they should or should not do.

Coping with the Stresses of Life

This category described the CRNs' experiences of that the women's recovery was influenced by their ability to cope with the stresses of life such as: concerns for themselves and returning to work, the fear of dying and not doing enough for the family. Furthermore, the CRNs described that women's recovery was affected by whether the woman had: a socially vulnerable situation with high demands, lack of finances, additional diseases and intensive treatments, as well as, whether they had experienced previous severe life events.

Regarding the feeling that life is threatened, it became clear that the women's recovery was affected by the fear of dying and of being inadequate. The women admitted to keeping a closer watch on their own physical and mental symptoms which contributed to them noticing everything that felt different within the body compared with before. "There are ... many who talk about ... that they have, indeed a fear that it will happen again." (n3). These thoughts came in the evening when it was quiet; therefore, sleeping habits could be adversely affected. Not infrequently, women had difficulty formulating their concerns into words and they avoided any discussions that they had been close to death. Instead, they had a need to dwell upon what had happened, to digest their new situation, in order to be able go forward in the recovery process. The women also had concerns about sexuality, and there were women who did not dare to have sexual intercourse as they were afraid that something could happen to the heart. The concern could also be about the fear of being sexually inactive and that inactivity and loss of desire would be permanent. Furthermore, it was also found that women were concerned about how the family would manage, now that they had suffered from ill health and could not cope with everyday tasks in the same way as before. "But it also seems like there is much concern for ... for the family ... usually the first ... you could say the first three or four months" (n1). Even the thought of returning to work concerned the women. Work could be experienced as difficult, both physically and mentally. "Obstacles to recovery, that is, I often have the feeling, those who are working, they will have the same amount of stress, they often say, as before. What will happen?" (n7). Many times, the women wished to have a longer period of sick leave in order to have time to their recovery.

Coping with stresses in life was affected by whether the women had an exposed social situation. The women's recovery was influenced by earlier completed traumatic life events, such as: abuse, divorce, serious illness within the family and death. "The fact that women have, quite often, experienced some rather severe life events which surface again when you have a heart attack. Yes, something from the past that you think you have got over, but it still pops up in your head, in connection with a heart attack. It can be anxiety, anguish; therefore, it becomes a tougher time before they recover." (n14). Furthermore, it was also found that there were women who stated that they already had a reduced work capacity, due to illness, resulting in a partial or full disability pension. This could be an additional burden that affected their recovery after an MI. Previous mental illness, depression and the onset of depression during the recovery phase also contributed to the fact that some women often declined to participate in CRP. "If you have experienced a breakdown in the past or suffered from depression; you are, of course, more vulnerable" (n14). Another factor that affected the women's recovery was which stage in life the women were in. The older women accepted, more easily, their illness, which they associated with age. They were not as easily affected by stress and often put more time and effort on lifestyle changes than younger women. Elderly women had more time and used it well, walking, as well as planning the purchase and cooking of healthy nutriatious food; which made the recovery easier. Younger women who worked and had young children did not have time to think and act this way.

Moreover, women who have had Coronary Artery Bypass Surgery [CABG] had more often a longer and tougher recovery period than women who underwent Percutaneous Coronary Intervention (PCI) or any other intervention, in connection to their MI. This was the reason why they felt fatigued and could not face up to current situations in the same way as the women who had undergone PCI. Even certain drugs and their side effects affect the women's daily life in a negative manner.

Furthermore, the women's social situation with poor finances and a lack of time, regardless of age, contributed to the fact that they did not have the opportunity to focus on themselves and their own situation. This is a reason why the women did not participate in CRP, which also affect their recovery process. "You can be on low income, perhaps a single parent, you are struggling financially and you become sick too. Yes, things like that." (n16). The women, intellectually, understood that CRP group meetings would be important to their recovery, but regardless of how hard they tried to plan their lives, there was a lack of time and money.

Wanting to Be Involved

This category described the CRNs' experiences of how the women's recovery was influenced by their interest and desire to know how they, themselves, affect their new life situation and make sure that they prioritized themselves.

The women's recovery was affected by their own interest of information with regard to the desire to know of risk and health factors, and understanding what exactly an MI was and what they should think about in the future. "The experience is that women are more interested in finding out, getting knowledge than men. They have more questions and concerns and it feels like they care more" (n4). They wanted to learn about the importance of diet and exercise in order to recover, how their prescribed drugs worked and about the body's normal physical and mental reactions. When the women had many questions concerning their new life situation, the new-found knowledge gave them a certain security. CRPs, which lasted for four to five weeks, gave them the opportunity to take in the knowledge gradually. The more information the women took in, the higher was their chance to recover.

Through the women's involvement in their care, the awareness of their own capacity was strengthened and led them not to see themselves only as "sick persons". They realized that they should be in charge of their own lives; therefore, changing their present situation into something better, which also led to a part of their recovery, "That you gain an insight into your illness and that you can accept this and get an understanding of what it implies" (n12). Likewise, when the women did group physical training, they became aware of what they managed without getting chest pain. This meant that they became less worried about straining themselves physically when they were on their own.

Regarding willingness to change, it was described that many women were interested in getting help and support in order to change their lifestyles, and therefore, prioritizing participation in various CRPs. The women, who went to CRPs, were motivated to learn new things regarding: diet, exercise, medications, and just what research revealed concerning health factors; which helped in the recovery. "Women change their diet, exercise habits and weight" (n15). Women were particularly keen to change their diet. Those, who found it easy to get started with physical training, recovered more quickly. Women, who relatively quickly went back to work, possibly only part-time, and those who tried to change their work situation to prevent negative stress, generally recovered more quickly than women who continued to be on sickleave. To effectively implement the changes in life was easier for the women who lived a somewhat carefree life when they fell ill and had a supportive family.

Relating to Yourself

This category highlighted the CRNs' experiences of that women's recovery was influenced by the way they related to themselves.

Regarding the feeling of shame and guilt, it emerged that there were women who expressed embarrassment over, both their smoking habits, and their overweight. They expressed an intention to stop smoking but they had relapsed and returned to their comforting smoking habits, which created a feeling of guilt. Not infrequently, they ended their participation in the CRPs which could affect their recovery adversely. Women also felt ashamed and guilty that they had fallen ill in the first place. "Women make themselves feel guilty ... I should not be doing this ... How did this happen? I should have foreseen" (n19). There were also women who felt guilty that their recovery took such a long time. These women needed to receive help to think through their life and work situation before they could continue their recovery.

The women's approach to life could affect their recovery. The women who had a positive attitude experienced, in general, less anxiety than those with a pessimistic attitude. "It is individual how you recover. It depends on what kind of personality you have ..." (n13). Even women, who had to have total control over their home situation, and thought that everything would work as it always had done, otherwise their world around them would collapse, usually had difficulty in recovering. Having to be perfect was a common feature of these women. These women did not allow themselves to be sick and prioritized others before themselves; therefore, hampering their own recovery. "When it comes to women so maybe ... they have to fix it by themselves ... They must be the strong one in the family in some way, they do not have time to be sick and it could be the career. Such things influence" (n12). Furthermore, there were women who had difficulty accepting their illness. They had difficulty in seeing themselves as sick and did not understand the onset of the illness; despite the many risk factors. The women with a more positive outlook on life changed their lifestyles when they now had got a second chance. They expressed that they had been given a new lease of life; thanks to: a deeper understanding of their first MI, an altered diet, new exercise habits, and their letting go of certain everyday demands. However, there were women who reported finding it challenging to relax their demands on themselves regarding the performance of everyday tasks.

DISCUSSION

The CRNs experienced that women's recovery after an MI was influenced by factors related both to their surroundings as well as by their own factors. The underlying meaning of the women's recovery can be described as: the transition process to the recovery to health. This transition process agrees with Schumacher and Meleis [34] interpretation of the concept "transition" as a profound inner process that influences the whole person and implies a new appreciation of life. The transition process to the recovery to health of the women, according to the CRNs' experiences in this study, depended on the women's ability to cope with the new life situation, as well as if they wanted to be involved in their care and were interested in increasing their knowledge of their illness and their treatment options. The transition process also depended on whether they had a positive approach to life and if the women had a good supportive context. This is in line with Meleis et al. [35] who describe that transition requires that a person has the ability to absorb new knowledge, change their behavior and thus the definition of self. It is important that CRNs are aware, that transition after an MI is a personal journey; so therefore, nursing should be person-centered in order to promote the women's unique recovery. Thus, CRNs should not put a time limit for the transition period because there are critical points and events which can affect the transition process [35]. An essential requirement for recovery for the women was access to support from their surroundings such as family, relatives, friends, and workplace as well as from other women with the same experiences. The women's recovery was, however, influenced negatively when they lacked support from their families. This indicates that it is important that families receive support from health care professionals in order to have the strength to give support to the ill person, which is confirmed by Ziegert [36]. The results even show that the women wanted counseling and support from the health care professionals during their recovery process. This is supported by Meleis et al. [35] who describes that confirmation, counseling and feedback are presumptions in order to get a successful transition to health. Earlier research has shown that the length of time of recovery for women, who received support from health care professionals throughout CRP, was shorter than for women who did not receive this support [37]. Therefore, it is essential that the CRNs support the women both in their clinical and personal recovery [14, 16, 17, and 19] and give the women the opportunity to an individually designed secondary prevention. Support from their surroundings, gives the women an opportunity to share their thoughts and gives people an understanding for the women's new situation. Women also need to share their thoughts with people with the same experience. The importance of peer support for women is well-known in research [38, 39] where women describe that peer support gives them valuable experiences when they have the opportunity to share their feelings with other women. This indicates that peer support is an essential intervention in order to promote the women's recovery. Thus, the CRNs are faced with the challenge of how to best organize and integrate peer support interventions into the usual health care.

The results show that the women's recovery was influenced by the fear of dying and the feeling of not being adequate, which is confirmed in earlier studies [12, 18, 19, 22, 40]. Fears of dying in our results show that the women felt anxiety and were afraid of suffering another MI. These findings correspond to earlier studies that indicate that women with MI can experience this worry and anxiety; especially during the first four months [12, 19, 22, 40, and 41]. Therefore, it is important that CRNs meet the women's needs and discuss their concerns so that the anxiety does not become pathological and negatively interfere their ability to recovery for a longer time. Furthermore, sexual concerns seem to induce worry and anxiety for the women. In Arenhall et al.'s study [42] male partners experienced that their partner was more fragile after an MI, which could result in male partners becoming more hesitant when it came to sexual intimacy. Research has shown that knowledge regarding sexuality after an MI is poor for both male and female patients. In fact, females have even lesser knowledge in comparison to males [43]. CRNs' knowledge and counseling, related to sexual issues, can play a key role: both in a supportive manner as well as impeding women to feel safe to talk about sexual concerns during their recovery process. Therefore, CRNs need both knowledge and specific education in order to provide counseling and support when addressing sexual issues [44-46].

Other factors that affected the women's ability to cope with their new life situation were whether the women had a socially vulnerable situation i.e. poor finances and lack of time. These factors, regardless of the women's age, contributed to the fact that they did not have the opportunity to focus on themselves and their own situation. This could be one of the reasons why the women do not participate in CRP, which is also underlined in others studies [12, 22]. Research has also shown that acute cardiac care, where both male and female patients from lower income groups did not undergo intervention as often as high earners; is a fact which affected their recovery [47]. Therefore, nurses must attach importance to being aware of socio-economic disparities when it comes to having access to both acute care and secondary prevention [48]. To address the barriers related to the access to secondary prevention, due to e.g. poor finances and place of residence, new innovative methods such as mHealth and eHealth need to be developed [49,50]. As peer support seemed to be important for women's recovery [38,39], support by telephone or internet from peers may provide sufficient support for women who for some reason are isolated and not able to attend CRPs at hospitals [51]. As research even has shown that eHealth has been suggested as a potential solution for health care professionals to connect to patients who not are able to participate to CRPs at hospitals, it is important that the nurses develop new innovative digital methods regarding eHealth for patients [52]. In addition, the development of digital CRPs present particular challenges as these services should be accessible to all patients in order to promote a more equitable care. This is important as low socio-economic status has been shown to be associated with increased heart disease incidence and mortality [24, 53].

The results show that women's recovery was influenced by their interest and desire of knowledge of how they, themselves, could affect their new life situation and how they prioritized themselves. If the women learned that they should not see themselves as ill and that they could take control of their lives by changing their situation into something better; the recovery process was affected positively. The individual recovery process is about learning to live with the changes that an MI entails [18,19,34], as well as a high degree of involvement in, and knowledge of, your own care. All this promotes a successful personal transition process [34,35]. Research shows that if patients are involved in their own recovery process it leads to better results of the treatment, and to a higher degree of satisfaction regarding the care [54]. It also shows how important it is that CRPs strive to care for the relationship with the patient, as research has found that the quality of the nurse-patient relationship is of great importance for the patient's involvement [55]. Many patients have expressed the wish to have more involvement in the MI-care; certainly more than they are offered at present [56]. Therefore, this requires that the nurses encourage the patient to be involved in her own process of recovery as well as nurses are aware of people's different needs for tailored information to gather knowledge [57,58].

Furthermore, the results show that CRNs experienced that the women expressed embarrassment and feelings of guilt over their lifestyle and that they had fallen ill. They expressed an intention to stop smoking but many relapsed into their comforting smoking habits, often ending their participation in the CRPs and then were left with a feeling of guilt that their recovery had taken such a long time. The result of guilt due to illness also corresponds to earlier research [40]. It is important that CRNs acknowledge the women's attempt to change their life situation and not to blame them if they do not succeed. If the women feel ashamed of their lifestyle and that they suffered a first MI, it will not lead to empowerment, but to the opposite. Results even show that women, who have to have total control over their home situation, usually experienced difficulty in recovering. The pressure of being perfect was a common feature of these women who did not allow themselves to be sick and put themselves last therefore, hampering their recovery. Research describes that it is common that women, during their recovery process, still continue to be responsible for much of the work at home; including keeping in contact with relatives and friends [28]. Schumacher and Meleis [34] describe that persons who can handle their new life situation regarding surroundings and who can manage the demands of the new situation promotes their transition more positively than those who cannot. Therefore, this suggests that the women's difficulties in letting go of the household chores to their partner does not depend on their partner's incompetence but because women want to keep their matriarchal gender role [28,29]. Worth noting is that CRNs, in this study, did not highlight women's stress as a factor affecting women's recovery, since earlier research shows that stress is a major factor for MI [12,38].

Methodological Consideration

In order to increase the variation in the material a quailtative content analysis was chosen [32]. The purposive selection of cardiac rehabilitation nurses, who have different experiences (age, different working experience in cardiac care, different hospitals) of female patients' recovery, increased the potential for achieving rich variation of the phenomenon under study, which strengthen the credibility but among the respondents there were no male CRN and this could limit the credibility of the study. Qualitative interviews with the same main question were deemed suitable for catch the aim. The follow-up questions allowed the CRNs to reflect and expand their views on the factors that influence female patients' recovery after their first MI. Another aspect of credibility in the present study was seeking agreement between the authors, how well the codes, categories and the theme covered the data. The authors worked both individually and together during the analysis process until a consensus was reached. Likewise, if the chosen quotations reflect the content of each category, it offers the reader an opportunity to determine the credibility of the study [32]. To establish dependability frequent discussions and a continuous dialogue among the authors took place so that our standpoints concerning differences and similarities of content were consistent over time [32]. The results could be of interest from the transferability perspective, as the CRNs' experiences can be useful in cardiac clinical nursing as a contribution to the ongoing discussion and improvement of nursing care connected with female patients' recovery after an MI [32].

CONCLUSIONS AND IMPLICATION

Cardiac rehabilitation nurses experienced that women's recovery after an MI was influenced by factors that were both related to their surroundings as well as by their own individual factors. The underlying meaning of women's recovery is characterized as the transition process for a recovery to health. This transition process is influenced by the women's ability to cope with the new life situation, if they wanted to be involved in their care, if they had a positive approach to life, as well as, if the women had a good supportive context. Our findings suggest that a focus on person-centered nursing would be beneficial in order to promote the every woman's personal and unique recovery after an MI. The present study also emphasizes the importance of how new innovative digital methods could be developed in order to face the barriers related to the access to CRP and increasing its accessibility, in order to meet different patients' needs. These findings also raise questions about the women's own attitudes and experiences of how their recovery is promoted in the MI care. Finally, the CRNs' experiences of factors influencing male patients' recovery after their first MI should be important to investigate.

REFERENCES

1. Gaziano, T.A., Bitton, A., Anand, S., Abrahams-Gessel, S. and Murphy, A. (2010) Growing epidemic of coronary heart disease in lowand middle-income countries. Current Problems in Cardiology, 35, 72-115. doi:10.1016/j.cpcardiol.2009.10.002
2. Wenger, N.K. (2003) Coronary heart disease: The female heart is vulnerable. Progress in Cardiovascular Diseases, 46, 199-229. doi:10.1016/j.pcad.2003.08.003
3. The National Board of Health and Welfare (2008) The National Board of Health and Welfare guidelines for cardiac care. Socialstyrelsen, Stockholm.
4. Johansson Sundler, A. (2008) Mitt hjärta, mitt liv: Kvinnors osäkra resa mot hälsa efter en hjärtinfarkt. My heart, my life: Women's uncertain health journey following a myocardial infarction. Växjö University Press, Växjö.

5. Perk, J., Guy De Backer, G., Gohlke, H., Graham, I., Zeljko Reiner, Z., Verschuren, M., et al. (2012) European guidelines on cardiovascular disease prevention in clinical practice (version 2012). European Heart Journal, 33, 1-77.
6. SWEDEHEART (2012) SWEDEHEART 2011 annual report. Matador Kommunikation, Uppsala.
7. World Health Organization (WHO) (2005) Preventing chronic diseases: A vital investment: WHO global report. World Health Organization, Geneva.
8. World Health Organization (WHO) (1993) Needs and action priorities in cardiac rehabilitation and secondary prevention in patients with coronary heart disease. WHO Regional Office for Europe, Geneva.
9. Piepoli, MF., Corrà, U., Benzer, W., Bjarnason-Wehrens, B., Dendale, P., Gaita D., et al. (2010) Secondary prevention through cardiac rehabilitation: from knowledge to implementation. A position paper from the cardiac rehabilitation section of the European association of cardiovascular prevention and rehabilitation. Cardiac rehabilitation section of the European association of cardiovascular prevention and rehabilitation. European Journal of Cardiovascular Prevention and Rehabilitation, 17, 1-17.doi:10.1097/HJR.0b013e3283313592
10. Jackson, L., Leclerc, J., Erskine, Y. and Linden, W. (2005) Getting the most out of cardiac rehabilitation: A review of referral and adherence predictors. Heart, 91, 10-14.doi:10.1136/hrt.2004.045559
11. De Feo, S., Tramarin, R., Ambrosetti, M., Riccio, C., Temporelli, P.L., Favretto, G., et al. (2012) Gender differences in cardiac rehabilitation programs from the Italian survey on cardiac rehabilitation (ISYDE-2008). International Journal of Cardiology, 160, 133-139.doi:10.1016/j.ijcard.2011.04.011
12. Sjöström-Strand, A., Ivarsson, B. and Sjöberg, T. (2011) Women's experience of a myocardial infarction: 5 years later. Scandinavian Journal of Caring Sciences, 25, 459- 466. doi:10.1111/j.1471-6712.2010.00849.x
13. Jackson, J.C., Hart, R.P., Gordon, S.M., Shintani, A., Truman, B., May, L., et al. (2003) Six-month neuropsychological outcome of medical intensive care unit patients. Critical Care Medicine, 31, 1226-1234. doi:10.1097/01.CCM.0000059996.30263.94

14. Slade, M. and Hayward M. (2007) Recovery, psychosis and psychiatry: Research is better than rhetoric. Acta Psychiatrica Scandinavica, 116, 81-83. doi:10.1111/j.1600-0447.2007.01047.x

15. Mancini, M.A., Hardiman, E.R. and Lawson, H.A. (2005) Making sense of it all: Consumer providers' theories about factors facilitating and impeding recovery from psychiatric disabilities. Psychiatric Rehabilitation Journal, 29, 48-55. doi:10.2975/29.2005.48.55

16. Anthony, W.A. (1993) Recovery from mental illness: The guiding vision of the mental health service system in the 1990s. Innovations & Research, 2, 17-24.

17. Deegan, P. (1996) Recovery as a journey of the heart. Psychiatric Rehabilitation Journal, 19, 91-97.

18. Tod, A. (2008) Exploring the meaning of recovery following myocardial infarction. Nursing Standard, 23, 35- 42.

19. Tobin, B. (2000) Getting back to normal: Women's recovery after a myocardial infarction. Canadian Journal of Cardiovascular Nursing, 11, 11-19.

20. Fridlund, B. (2002) The role of the nurse in cardiac rehabilitations programmes. European Journal of Cardiovascular Nursing, 1, 15-18. doi:10.1016/S1474-5151(01)00017-2

21. Hedback, B., Perk, J. and Wodlin, P. (1993) Long-term reduction of cardiac mortality after myocardial infarction: 10-year results of a comprehensive rehabilitation programme. European Heart Journal, 14, 831-835. doi:10.1093/eurheartj/14.6.831

22. Kerr, E.E. and Fothergill-Bourbonnais, F. (2002) The recovery mosaic: Older women's lived experiences after a myocardial infarction. Heart & Lung, 31, 355-367.doi:10.1067/mhl.2002.127939

23. Rankin, S.H. (2002) Women recovering from acute myocardial infarction: Psychosocial and physical functioning outcomes for 12 months after acute myocardial infarction. Heart & Lung, 31, 399-410. doi:10.1067/mhl.2002.129447

24. McSweeney, J.C. and Coon, S. (2004) Women's inhibittors and facilitators associated with making behavioral changes after myocardial infarction. Medical-Surgical Nursing, 13, 49-56.

25. Svedlund, M. and Danielson, E. (2004) Myocardial infarction: Narrations by afflicted women and their partners of lived experiences in daily life following an acute myocardial infarction. Journal of Clinical Nursing, 13, 438- 446. doi:10.1111/j.1365-2702.2004.00915.x
26. Worrall-Carter, L., Jones, T. and Driscoll, A. (2005) The experiences and adjustments of women following their first acute myocardial infarction. Contemporary nurse. A Journal for the Australian Nursing Profession, 19, 211- 221.
27. Day, W. and Batten, L. (2006) Cardiac rehabilitation for women: One size does not fit all. Australian Journal of Advanced Nursing, 24, 21-26.
28. Hildingh, C., Fridlund, B. and Lidell, E. (2007) Women's experiences of recovery after myocardial infarction: A meta-synthesis. Heart & Lung, 36, 410-417.
29. Eriksson, M., Asplund, K. and Svedlund, M. (2009) Patients' and their partners' experiences of returning home after hospital discharge following acute myocardial infarction. Journal of Cardiovascular Nursing, 8, 267-273.doi:10.1016/j.ejcnurse.2009.03.003
30. Stevens, S. and Thomas, S.P. (2012) Recovery of midlife women from myocardial infarction. Health Care for Women International Health, 33, 1096-1113.doi:10.1080/07399332.2012.684815
31. Downe-Wamboldt, B. (1992) Content analysis: Method, applications, and issues. Health Care for Women International, 2, 313-321. doi:10.1080/07399339209516006
32. Graneheim, U.H. and Lundman, B. (2004) Qualitative content analysis in nursing research: Concepts, procedures and measures to achieve trustworthiness. Nurse Education Today, 24, 105-112. doi:10.1016/j.nedt.2003.10.001
33. World Medical Association (2008) Declaration of Helsinki—Ethical principles for medical research involving human subjects. The World Medical Association (WMA), Seoul.
34. Schumacher, K.L. and Meleis, A.I. (1994) Transitions: A central concept in nursing. The Journal of Nursing Scholarship, 26, 119-127. doi:10.1111/j.1547-5069.1994.tb00929.x

35. Meleis, A.I., Sawyer, L.M., Im, E., Messias, D.K. and Shumacher, K. (2000) Experiencing transitions: An emerging middle-range theory. Journal of Advanced Nursing, 23, 12-28.

36. Ziegert, K. (2011) Maintaining families' well-being in everyday life. International Journal of Qualitative Studies on Health and Well-Being, 6.

37. Wieslander, I., Baigi, A., Turesson, C. and Fridlund, B. (2005) Women's social support and social network after their first myocardial infarction: A 4-year follow-up with focus on cardiac rehabilitation. European Journal of Cardiovascular Nursing, 4, 278-285.doi:10.1016/j.ejcnurse.2005.06.004

38. Sjöström-Strand, A. and Fridlund, B. (2006) Women's descriptions of coping with stress at the time of and after a myocardial infarction: A phenomenographic analysis. Canadian Journal of Cardiovascular Nursing, 16, 5-12.

39. White, J., Hunter, M. and Holttum, S. (2007) How do women experience myocardial infarction? A qualitative exploration of illness perceptions, adjustment and coping. Psychology Health & Medicine, 12, 278-288. doi:10.1080/13548500600971288

40. Svedlund, M., Danielson, E. and Norberg, A. (2001) Women's narratives during the acute phase of their myocardial infarction. Journal of Advanced Nursing, 35, 197- 205.doi:10.1046/j.1365-2648.2001.01837.x

41. Alsén, P., Brink, E., Persson, L.O., Brändström, Y. and Karlson, B.W. (2010) Illness perceptions after myocardial infarction: Relations to fatigue, emotional distress, and health-related quality of life. Journal of Cardiovascular Nursing, 25, E1-E10.

42. Arenhall, E., Kristofferzon, M.L., Fridlund, B., Malm, D. and Nilsson, U. (2011) The male partners' experiences of the intimate relationships after a first myocardial infarction. European Journal of Cardiovascular Nursing, 10, 108-114.doi:10.1016/j.ejcnurse.2010.05.003

43. Nilsson, U.G., Svedberg, P., Fridlund, B., Alm-Roijer, C., Thylén I. and SAMMI-Study Group (2012) Sex knowledge in males and females recovering from a myocardial infarction: A brief communication. Clinical Nursing Research, 21, 486-494. doi:10.1177/1054773812437241

44. Jaarsma, T., Strömberg, A., Fridlund, B., De Geest, S., Mårtensson, J., Moons, P., et al. (2010) Sexual counselling of cardiac patients: Nurses' perception of practice, responsibility and confidence. European Journal of Cardiovascular Nursing, 9, 24-29. doi:10.1016/j.ejcnurse.2009.11.003

45. Jaarsma, T., Steinke, E.E. and Gianotten, W.L. (2010) Sexual problems in cardiac patients: How to assess, when to refer. Journal of Cardiovascular Nursing, 25, 159-164.

46. Steinke, E.E. (2010) Sexual dysfunction in women with cardiovascular disease: What do we know? Journal of Cardiovascular Nursing, 25, 151-158.

47. Rosvall, M., Chaix, B., Lynch, J., Lindström, M. and Merlo, J. (2008) The association between socioeconomic position, use of revascularization procedures and fiveyear survival after recovery from acute myocardial infarction. BMC Public Health, 1, 44.doi:10.1186/1471-2458-8-44

48. Haglund, B., Köster, M., Nilsson, T. and Rosén, M. (2004) Inequality in access to coronary revascularization in Sweden. Scandinavian Cardiovascular Journal, 38, 334-339. doi:10.1080/14017430410021516

49. Varnfield, M., Karunanithi, M.K., Särelä, A., Garcia, E., Fairfull, A., Oldenburg, B.F., et al. (2011) Uptake of a technology-assisted home-care cardiac rehabilitation program. The Medical Journal of Australia, 194, S15-S19.

50. Pfaeffli, L., Maddison, R., Whittaker, R., Stewart, R., Kerr, A., Jiang, Y., et al. (2012) A Health cardiac rehabilitation exercise intervention: findings from content development studies. BMC Cardiovascular Disorders, 12, 36. doi:10.1186/1471-2261-12-36

51. Parry, M.J., Watt-Watson, J., Hodnett, E., Tranmer, J., Dennis, C.L. and Brooks, D. (2009) Cardiac Home Education and Support Trial (CHEST): A pilot study. Canadian Journal of Cardiology, 25, 393-398. doi:10.1016/S0828-282X(09)70531-8

52. Vandelanotte, C., Dwyer, T., Van Itallie, A., Hanley, C. and Mummery, W.K. (2010) The development of an internet-based outpatient cardiac rehabilitation intervenetion: A Delphi study. BMC Cardiovascular Disorders, 10, 27. doi:10.1186/1471-2261-10-27

53. Pollitt, R.A., Rose, K.M. and Kaufman, J.S. (2005) Evaluating the evidence for models of life course socioeconomic factors and cardiovascular outcomes: A systematic review. BMC Public Health, 5, 7-13.

54. Williams, G.C., Freedman, Z.R. and Deci, E.L. (1998) Supporting autonomy to motivate patients with diabetes for glucose control. Diabetes Care, 21, 1644-1651.doi:10.2337/diacare.21.10.1644

55. Sahlsten, M.J., Larsson, I.E., Sjöström, B., Lindencrona, C.S. and Plos, K.A. (2007) Patient participation in nursing care: Towards a concept clarification from a nurse perspective. Journal of Clinical Nursing, 16, 630-637. doi:10.1111/j.1365-2702.2006.01660.x

56. Arnetz, J.E. and Arnetz, B.B. (2009) Gender differences in patient perceptions of involvement in myocardial infarction care. European Journal of Cardiovascular Nursing, 8, 174-181. doi:10.1016/j.ejcnurse.2008.11.002

57. Young, L.E. and Murray, J. (2011) Patients' perception of their experience of primary percutaneous intervention for ST segment elevation myocardial infarction. Canadian Journal of Cardiovascular Nursing, 21, 20-30.

58. Sherrod, M., McIntire Sherrod, N., Tate Spitzer, M. and Cheek, D. (2012) Prevention of heart disease in women: Considerable challenges remain. Open Journal of Nursing, 2, 176-180. doi:10.4236/ojn.2012.23027

Chapter 7

Monitoring Geologic Hazards and Vegetation Recovery in the Wenchuan Earthquake Region using Aerial Photography

Zhenwang Li[1, 2], Quanjun Jiao[1], Liangyun Liu[1], Huan Tang[1, 2], and Tong Liu[1]

[1]Institute of Remote Sensing and Digital Earth, Chinese Academy of Sciences, Beijing 100094

[2]Institute of Agricultural Resources and Regional Planning, Chinese Academy of Agricultural Sciences, Beijing 100094, China

ABSTRACT

On 12 May 2008, the 8.0-magnitude Wenchuan earthquake occurred in Sichuan Province, China, triggering thousands of landslides, debris flows, and barrier lakes, leading to a substantial loss of life and damage to the local environment and infrastructure. This study aimed to monitor the status of geologic hazards and vegetation recovery in a post-earthquake disaster area using high-resolution aerial photography from 2008 to 2011, acquired from the Center for Earth Observation and Digital Earth (CEODE), Chinese Academy of Sciences. The distribution and range of hazards were identified in 15 large, representative geologic hazard areas triggered by the Wenchuan earthquake. After conducting an overlay analysis, the variations of these hazards between successive years were analyzed to reflect the geologic hazard development and vegetation recovery. The results showed that in the first year after the Wenchuan earthquake, debris flows occurred frequently with high intensity. Resultantly, with the source material becoming less available and the slope structure stabilizing, the intensity and frequency of debris flows gradually decreased with time. The development rate of debris flows between 2008 and 2011 was 3% per year. The lithology played a dominant role in the formation of debris flows, and the topography and hazard size in the earthquake affected area also had an influence on the debris flow development process. Meanwhile, the overall geologic hazard area decreased at 12% per year, and the vegetation recovery on the landslide mass was 15% to 20% per year between 2008 and 2011. The outcomes of this study provide supporting data for ecological recovery as well as debris flow control and prevention projects in hazard-prone areas.

INTRODUCTION

On 12 May 2008, the 8.0-magnitude Wenchuan earthquake occurred in Sichuan Province, China. Its epicenter was located at Yingxiu Town, Wenchuan County, in the east of the Longmenshan Fault Zone on the eastern fringe of the Tibetan Plateau. The Wenchuan earthquake caused substantial casualties and financial losses. In response to the massive energy this earthquake released, thousands of geologic failures were triggered such as landslides and debris flows, in addition to the formation of barrier lakes [1,2]. These failures caused enormous damage to the local environment and ecosystem [3], and destabilized numerous hillsides and solid masses, increasing the probability of geologic disasters [4]. Particularly in the mountainous areas, the damage caused by earthquake-triggered secondary geologic failures may have been more severe than the immediate damage caused by the earthquake itself [5].

After the earthquake, timely estimating the development of geo-hazards and disaster conditions through field investigation was necessary for hazards assessment and decision making. However, damaged roads, unstable hillsides, inclement weather, and frequent aftershocks made it difficult to perform field work and acquire reliable information about the areas. Remote sensing is an effective means for the timely monitoring and quantitative assessment of devastated areas [6]. Remote sensing data can be acquired at broad spatial and temporal scales, and provide useful information for rapid damage assessment at an early stage when field surveys of a large area are difficult [7]. Remote sensing data have been available for approximately one century and have proven useful in natural-hazard investigation and management in response to a number of natural disasters [8,9,10,11,12,13,14]. Today, as remote sensing technology develops toward being multiplatform, multispectral, high-resolution, and multi temporal, it has become more convenient to use in hazard assessment and rapid monitoring [15].

Much research has been conducted on the Wenchuan earthquake using optical remote sensing data [16,17,18] and multi-mode radar data [11,19,20], focusing on disaster monitoring [21,22,23,24,25], hazard assessment [26,27,28,29], mitigation [30,31] and vegetation damage assessment [3,32]. Chigira *et al.* [33] discussed the distribution and characteristics of seismically triggered landslides from the Wenchuan earthquake using satellite images. Huang and Li [34] studied the distribution of geo-hazards triggered by the earthquake, and identified a total of 11,308 geo-hazards in sixteen seriously damaged counties based on field investigations, air photos, and satellite imaging data. Ge *et al.* [35] estimated the damage to vegetation caused by the secondary hazards of the Wenchuan earthquake through SPOT and aerial images. The authors found that farmland and grassland were less damaged than forestland because of the steep slopes and that the damage to the vegetation first decreased and then increased with increasing distance from the three main faults of the Longmenshan fault zone. Liu *et al.* [36] evaluated the vegetation degradation caused by the earthquake and its recovery two months after the disaster from the analysis of MODIS Gross Primary Productivity (GPP) time series products and other ancillary GIS data. Xu and Lu [37] summarized a meta-synthesis pattern of the post-disaster recovery and reconstruction by comparing studies of 14 earthquakes in the last 100 years. The authors found that remarkable achievements have been made with respect to post-Wenchuan earthquake recovery and reconstruction and that these post-disaster recovery and reconstruction programs can be used as a foundation for future earthquake recovery.

Moreover, dynamic monitoring of geologic hazards triggered by the earthquake and aftershocks in successive years using remote sensing data is necessary and important to uncover the mechanism and influence factors in the geological processes. Because of the limited availability of images at a very high spatial resolution (suitable for detecting geologic hazards) acquired periodically over the same area, long-term careful monitoring of post-earthquake disaster development and vegetation recovery through images are rarely conducted. In our research, satellite imagery from before the earthquake and four epochs of high-resolution airborne optical images after the Wenchuan earthquake were utilized to monitor and assess the development of secondary geologic hazards (refers to the geologic hazards triggered by the Wenchuan earthquake or aftershocks in this paper) by focusing on 15 representative large-scale geologic hazard sites in the affected areas. The condition of the vegetation recovery was also analyzed in the context of the landslides and debris flows, and variation rate was displayed to illustrate their development at different stages after the Wenchuan earthquake. Besides, geological and topographic factors were also analyzed to discover the hazard development and recovery trajectories. The results provide scientific data of the vegetation recovery and hazard development, which can be used for prevention and reconstruction practices in the Wenchuan earthquake area.

STUDY AREA

The Wenchuan earthquake occurred along the Longmenshan fault, a thrust structure along the border of the Indo-Australian Plate and Eurasian Plate. Seismic activities were concentrated along its mid-fracture (known as the Yingxiu-Beichuan fracture), causing substantial loss of life and damage to property, infrastructure, and the environment. Our study areas were the focus of the overlapping flight zones covered by airborne optical images acquired between 2008 and 2011 (see

Figure 1). This area covered most of the counties that were severely damaged by the Wenchuan earthquake, totaling approximately 20,000 km^2. These counties and cities include Beichuan County, Wenchuan County, Mianzhu City, Shifang City, Dujiangyan City, Pengzhou City, Mianyang City, Jiangyou City, and Pingwu County. The region is characterized by mountainous topography where the elevation ranges from 500 m to over 6000 m. The complex geologic structure of the study area makes it highly susceptible to landslides and debris flows.

DATA AND METHODS

Data Acquisition

Immediately after the Wenchuan earthquake, an emergency project, Monitoring and Assessment of the Wenchuan Earthquake Disasters by Remote Sensing, was initiated by the Chinese Academy of Sciences (CAS). A large number of remote sensing images of various types were acquired between 15 and 28 May 2008, for disaster assessment and rapid monitoring and evaluation of the severely damaged areas. The spatial resolution of these images ranged from 0.5 to 0.8 m depending on the flight altitude. During the subsequent years, airborne optical remote sensing images covering most of the heavily affected areas were again acquired in 2009, 2010, and 2011 to monitor the process of environmental restoration and reconstruction [38]. In our study, we used airborne optical images from 2008 to 2011 acquired from the Center for Earth Observation and Digital Earth (CEODE), CAS, to analyze the hazards development and ecological recovery in the earthquake-affected areas. Table 1 shows the flight task details.

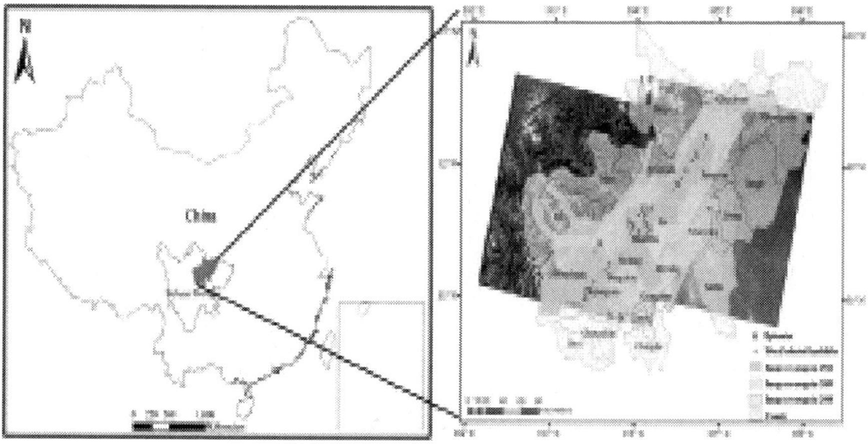

Figure 1: Airborne imagery across the earthquake region (background: Landsat TM image in 2007).

Table 1: Remote sensing data acquired over the earthquake-damaged area

Flight Year	Flight Data	Sensor	Image Resolution	Number of Images	Flight Area (km²)
2008	15 May–28 May	ADS40	0.5–0.7 m/2 m	99	36,744
2009	16 May–03 June	ADS80	0.5 m/5 m	108	33,266
2010	18 April–04 May	UCXp	0.3 m/4 m	78	Approximately 10,000
2011	17 May–07 June	UCXp	0.3 m/4 m	57	20,125

Note: the number before "/" in the "image resolution" column refers to the original spatial resolution, the number after "/" refers to the spatial resolution after resizing.

The spectral bands of the acquired images include blue, green, and red. The airborne sensors had a high spatial resolution, ranging from 0.3 to 0.7 m. However, the large data size and non-uniform image resolution made the images difficult to manage, thus after acquiring the images, CEODE resized the images to a uniform resolution. In addition, a systematic orthorectification of all the airborne images was conducted using the 25-m DEM data interpolated from 1:50,000 scale topographic maps [38], except for the year of 2010 because of the lack of Position and Orientation System (POS) data.

Auxiliary data used in this study included a Sichuan administrative vector map; 30-m resolution Digital Elevation Model (DEM) data (scene ID: srtm_57_06 and srtm_58_06) obtained by the Shuttle Radar Topography Mission (SRTM) and acquired for this study from the Land Cover Facility, University of Maryland; the China geological map, produced by the China Geological Survey, to acquire the lithology data; and 30-m resolution Landsat TM images from 2007 (entity ID: LE71290382007262PFS00 and LE71300382007141SGS00) acquired from the United States Geological Survey to reflect the land cover before the earthquake.

Method

Region Selection

A landslide is a downslope movement of rock or soil occurring on the surface of a rupture in which much of the material often moves as a coherent or semi-coherent mass with little internal deformation [39, 40]. Landslides triggered by the earthquake often showed the following characteristics on remote sensing images: vegetation was disturbed, the deposits could be clearly identified at the foot of the slope along roads or rivers, and the interrupted region showed a bright white or dark brown contrast compared with the surrounding slopes. Debris flow is a common type of fast-moving mass movement in which loose soil, rock, and sometimes organic matter combine with water to form a slurry that flows downslope [41]. Debris flows usually start on steep hillsides as liquefied shallow landslides, and continue flowing down and through channels, accumulating water, sand, mud, trees, and other material. When the flows reach canyon mouths or flatter ground, the debris spreads over a broad area [42]. On remote sensing images, debris flows are usually characterized as an obvious flow channel and forms a large alluvial fan at the foot of the slope (Figure 2b).

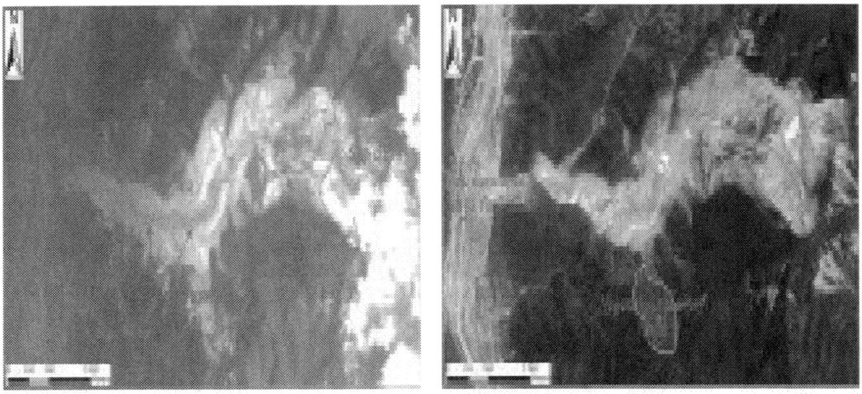

(a) (b)

Figure 2: Airborne remote sensing images of Wenjiagou, Qingping Town, Mianzhu City (site No. 1) taken on 23 May 2008 (a) and 7 July 2011 (b).

After the earthquake, aftershocks and heavy rain led to thousands of secondary geologic hazards, including landslides, debris flows, and barrier lakes. To illustrate the changes representing environmental restoration and human settlement reconstruction, representative study areas were selected. In this study, we first used a visual interpretation method to select large-scale landslides by comparing a Landsat TM image from 2007 and airborne optical images from 2008 and finally selected 15 geologic hazard sites distributed throughout the severely damaged areas (red points in Figure 1) with high quality of minimal or no cloud contamination in most of the flight year above the site. Each hazard site contained several landslides, with the locations and attributes shown in Table 2. Then, the corresponding images covering every hazard site were selected from 2008, 2009, and 2011 airborne optical images. Because of the poor data quality of the images from 2009 of the No. 14 site and from 2008 of the No. 15 site, the analyses of these two sites' secondary geologic hazard development were based on images from 2008, 2010, and 2011 and from 2009, 2010, and 2011, respectively.

Table 2: Statistics for the 15 geologic hazard s tes

No.	Location	Image Date			Number of Landslides	Total Area in 2008 (ha)
		2008	2009	2011		
1	Wenjiagou, Qingping Town, Mianzhu City	23 May	03 June	07 June	1	260.33
2	Chenjiaba, Beichuan County	28 May	18 May	31 May	23	631.63
3	East Yingxiu Town, Wenchuan County	23 May	16 May	07 June	8	121.68
4	Hanwang Town, Mianzhu City	19 May	16 May	25 May	11	113.30
5	Tianchi Town, Mianzhu City	23 May	16 May	25 May	12	148.18
6	Shawan, Nanba Town, Pingwu County	28 May	18 May	31 May	3	48.49
7	Yingxiu Town, Wenchuan County	23 May	03 June	28 May	14	664.73
8	`Jinhelinkuang, Mianzhu City	23 May	03 June	07 June	12	616.85
9	Jiuding Mountain, Bailongchi, Mianzhu City	23 May	16 May	25 May	2	110.33
10	Tangjiashan, Beichuan County	27 May	18 May	31 May	27	487.14
11	Guanzhipu, Beichuan County	28 May	18 May	31 May	13	155.32
12	Jinchi Town, Wenchuan County	24 May	21 May	28 May	5	191.11

13	Gaochuan County, Jinyang City	23 May	16 May	25 May	6	373.82
14 *	Changheba, Mianzhu City	23 May	miss	25 May	2	141.90
15 *	Daguangbao Group, An County	miss	16 May	25 May	4	959.59

Note: "*" used here in the No. 14 and 15 sites indicates that the missing data are infilled with data from 28 April 2010 and 27 April 2010, respectively.

Geometric Registration

Although the acquired aerial images were subjected to geometric and orthorectification corrections, there was still position deviation to a certain extent because of the complex terrain. After comprehensive consideration of the image quality and position error compared with the Landsat TM image, we selected airborne images from 2009 as our reference images, and selected 100 to 150 control points at each hazard site to complete the geometric registration of the images from 2008, 2010, and 2011. In addition, the China geological map was projected in UTM coordinates (WGS84 datum), the same as Landsat image.

Detection of Geologic Hazards

The mountainous parts of the study area had high vegetation cover prior to the earthquake, spatially covering up to 80% of the entire region [43]. Earthquake-induced geologic hazards such as landslides and debris flows lead to large-scale vegetation interruption and some feature changes in the landscape, including the color, texture, and shape. These distinctive features help with identification of failure areas in remote sensing images. Figure 2 presents the airborne remote sensing images of landslides and debris flows in Qingping Town in 2008 and 2011.

After geometric registration and post-processing, the airborne optical images were used to construct maps in true color and were analyzed for hazard damage monitoring. Using an interactive interpretation method between pre-earthquake Landsat TM images and post-earthquake optical airborne images, the 15 selected geologic hazard areas from 2008, 2009, and 2011 were carefully interpreted in the ArcGIS platform through polygons drawn based on the specific features and differences described in Section 3.2.1.

Based on the interpretation results, the slope failure and vegetation recovery areas were obtained by performing an overlay analysis with the previous year's failure areas. After eliminating newly developed landslide mass areas and areas with registration errors on the overlay analysis results through artificial screening, every site's newly developed debris flow areas (the newly emerged hazard area caused by debris flows) were obtained. Then, the vegetation recovery areas based on the previous failure areas were calculated with Equation (1). The rates of vegetation recovery, debris flow development, and geologic hazard variation were calculated with Equation (2).

$$na = oa + p - er \tag{1}$$

$$rate = \frac{\text{var}}{oa * years} \tag{2}$$

In Equation (1), *na* refers to the bare failure area after the vegetation recovered; *oa* is the bare failure area prior to vegetation recovery; *p* is the newly developed debris flow area; and *er* is the vegetation recovery area based on the previous failure region. In Equation (2), *rate* refers to the variation rate of the three objects, *years* refers to the years apart of analyzed variables, and var refers to the three analysis objects. In this paper, var equals (*oa* − *na*) if the object is geologic hazard variation, *p* if the object is debris flow development, and *er* if the object is vegetation recovery.

Field Investigation

To verify the interpretation results and identify the reason for various hazard developments, *in-situ* surveys were conducted from 6 to 13 May 2012 in several large slope movements. The field investigation covered most of the selected hazard sites (except for No. 8, No. 9 and No. 14) that were relatively easy to access, no equipment was used during the survey, and basic conditions in each site were recorded manually. The elements recorded included the soil structure (mainly particle composition), human activity (whether good for hazard prevention and recovery), and vegetation recovery conditions (yes or no) and patterns (naturally or artificially). Because of the complex conditions in each site for the large area and combined hazard group, the survey results were not listed here and were used in the following section to explain the remote sensing based results. Several example *in-situ* photos are shown in Figure 3.

RESULTS

Multi-Temporal Detecting Geological Hazards

The 2008, 2009, and 2011 secondary geologic hazard areas, as determined by the above analysis, are shown in Table 3. This table shows that during the first year after the earthquake, the secondary geologic hazard development differed greatly among sites, with five sites experiencing an increased area and ten sites experiencing a decreased area. In 2011, however, almost all of the sites, except No. 14 and No. 15, experienced a decrease in hazard area, although at varying rates. The changes of geological hazards between 2008 and 2011 of several hazards are shown in Figure 4. From the monitoring results, the variations of geological hazards could be classified into four cases. The first case was for a continuous decrease in hazard area from 2008 to 2011 that included No. 2, No. 4, No. 5, No. 6, No. 7, No. 8, No. 12 and No. 13, accounting for 53.3% of all sites. The second case was for an increase in hazard area from 2008 to 2009, and a decreased hazard area in the next two years, such as the No. 3, No. 10, No. 11 and No. 15 sites. The third case was for a decreased area from 2008 to 2009 and an increase in hazard area in the next two years, such as the No. 1 and No. 9 sites. The last case was for a continuous increase in hazard area from 2008 to 2011, which pertained only to site No. 14. Nonetheless, from 2008 to 2011, the overall geologic hazard area decreased as vegetation growth rate exceed that of debris flow development, and the geologic hazards variation rate between 2008 and 2011 was approximately 12%, which was higher than the rate of the first year after the earthquake.

Figure 3: Photos taken at geologic hazard sites during the field investigation in May 2012. (**a**) Vegetation recovery in Yingxiu Town, Wenchuan County (site No. 7); (**b**) Old Beichuan Town, Beichuan County (site No. 10); (**c**) The huge failure in Tangjiashan, Beichuan County (site No. 10); (**d**) Vegetation recovery on the failure slope in Qingping Town, Mianzhu City (site No. 1).

Table 3: Each site's geologic hazard area and variation rate

No.	Geologic Hazard Area (ha)			**Geologic Hazard Variation Area (ha)**			**Geologic Hazard Variation Rate (per year)**		
	2008	**2009**	**2011**	**2008–2009**	**2009–2011**	**2008–2011**	**2008–2009**	**2009–2011**	**2008–2011**
1	260.33	232.70	235.03	−27.63	2.34	−25.29	0.11	−0.01	0.03
2	631.63	626.37	323.61	−5.26	−302.76	−308.02	0.01	0.24	0.16
3	121.68	122.09	107.11	0.41	−14.98	−14.57	−0.003	0.06	0.04
4	113.30	86.60	53.21	−26.70	−33.39	−60.09	0.24	0.19	0.18
5	148.18	129.20	88.81	−18.98	−40.39	−59.37	0.13	0.16	0.13
6	48.49	45.58	39.74	−2.91	−5.84	−8.75	0.06	0.06	0.06
7	664.73	454.83	265.57	−209.89	−189.26	−399.16	0.32	0.21	0.20
8	616.85	602.69	463.49	−14.16	−139.20	−153.36	0.02	0.12	0.08
9	110.33	102.74	105.83	−7.59	3.09	−4.51	0.07	−0.02	0.01
10	487.14	516.83	351.75	29.70	−165.08	−135.39	−0.06	0.16	0.09
11	155.32	163.12	109.39	7.80	−53.73	−45.93	−0.05	0.16	0.10
12	191.11	154.99	111.55	−36.12	−43.44	−79.56	0.19	0.14	0.14
13	373.82	230.04	224.42	−143.77	−5.63	−149.40	0.38	0.01	0.13
14	141.90	153.84 (2010)	164.60	11.94	10.76	22.70	−0.04	−0.07	−0.05
15	959.59 (2009)	972.15 (2010)	951.05	12.56	−8.53	12.56	−0.01	0.01	−0.01
Total	3922.91	3467.78	2479.51	−455.10	−988.27	−1443.40	0.116	0.142	0.123

Note: In the column of "geologic hazards variation rate", positive numbers represent an decrease in hazard area; negative numbers represent a increase in hazard area; with regard to the No. 14 and No. 15 sites, the columns "2008–2009," "2008–2011," and "2009–2011" correspond to "2008–2010," "2009–2010," and "2010–2011," respectively, the "Total" row is the accumulation of the first thirteen hazards sites, and does not include the last two.

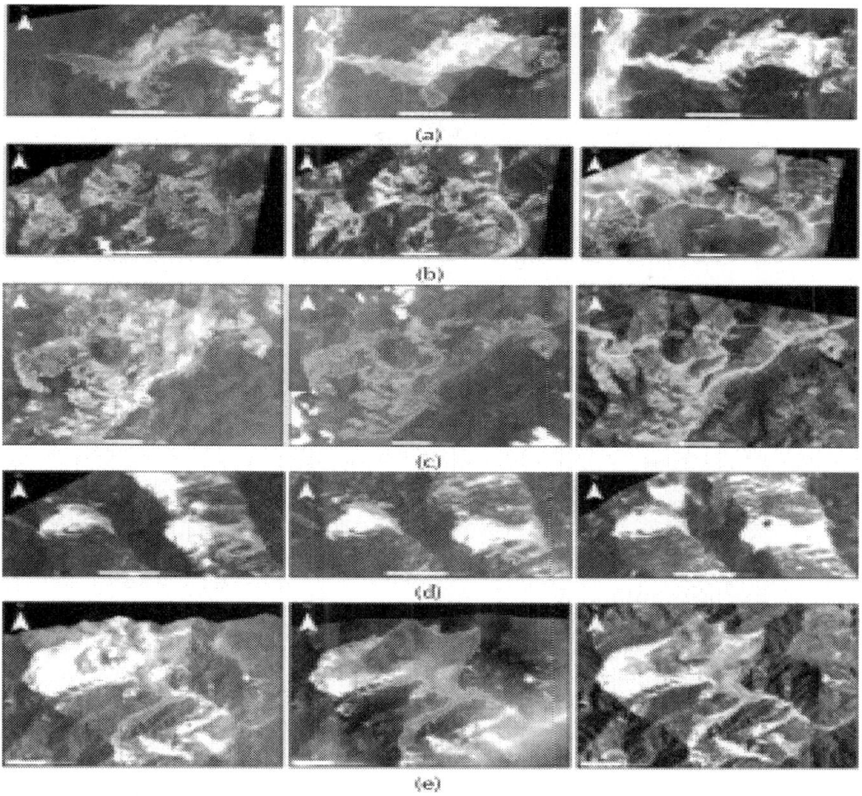

Figure 4: Airborne remote sensing images of several hazards in 2008, 2009, and 2011, respectively (from left to right). (**a**) The big landslide and debris flow of Wenjiagou, Qingping Town, Mianzhu City (site No. 1); (**b**) The big landslide group happened at Yingxiu Town, Wenchuan County (site No. 7); (**c**) The change of hazard happened at Tangjiashan, Beichuan County (site No. 10); (**d**) The change of hazard happened at Changheba, Mianzhu City (site No. 14); (**e**) The change of Daguangbao hazard group, AnCounty (site No. 15).

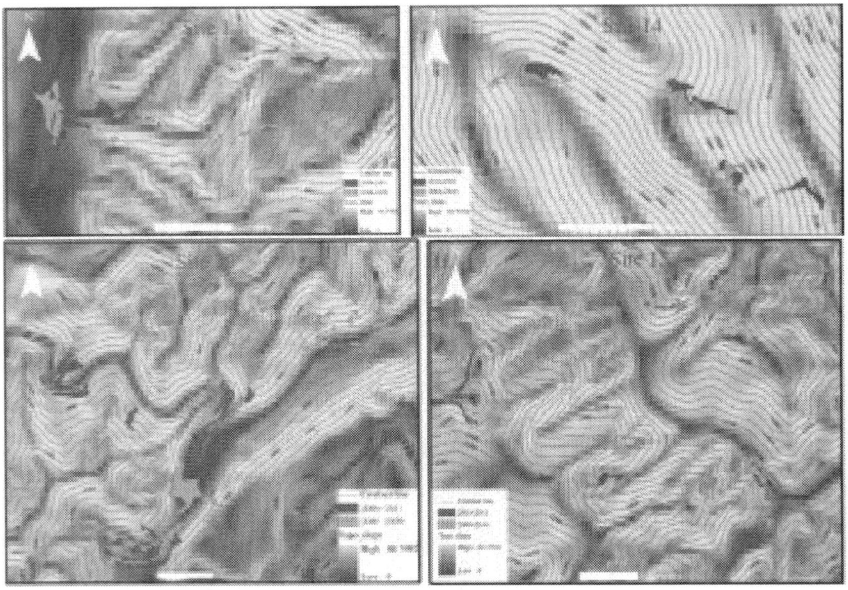

Figure 5: Changes in the debris flow regions of site 1, 10, 14, 15 from 2008 to 2009 and from 2009 to 2011.

Multi-Temporal Detecting Debris Flows

After acquiring the geologic hazard areas from 2008, 2009, and 2011, the newly developed debris flow regions were obtained utilizing an overlay analysis between successive years. Table 4 presents the development of debris flows from 2008 to 2009 and from 2009 to 2011, and debris flow variation region of several typical sites were illustrated in Figure 5. During the first year after the earthquake, the activity of debris flows was particularly severe, causing the development of many new hazards areas. After two years, the activity gradually decreased but was still occurring. The developed areas and rates are shown in Table 4. From 2008 to 2009, the newly developed debris flow areas reached 402.05 ha, which was far larger than the average area that developed in the next two years, and the development rate was almost triple that between 2009 and 2011. In general, the average development rate in these three years for the entire region was 3%, which is in accordance with Chen's study [44].

Table 4: Each site's newly developed debris flow areas and rates

No.	Newly Developed Debris Flow Area (ha)			**Debris Flow Development Rate (per year)**		
	2008–2009	**2009–2011**	**2008–2011**	**2008–2009**	**2009–2011**	**2008–2011**
1	23.46	49.22	58.45	0.09	0.11	0.07
2	147.89	78.78	104.52	0.23	0.06	0.06
3	15.64	29.19	30.42	0.13	0.12	0.08
4	15.69	0.00	7.00	0.14	0.00	0.02
5	24.01	7.98	14.29	0.15	0.03	0.03
6	9.07	2.98	4.75	0.19	0.03	0.03
7	0.00	0.00	0.00	0.00	0.00	0.00
8	35.39	0.00	17.63	0.06	0.00	0.01
9	3.11	7.79	7.21	0.03	0.04	0.02
10	84.75	45.93	63.13	0.17	0.04	0.04
11	31.73	0.00	19.58	0.20	0.00	0.04
12	0.00	7.26	0.00	0.00	0.02	0.00
13	11.31	6.94	15.14	0.03	0.02	0.01
14	20.24 (2008–2010)	13.88 (2010–2011)	25.37 (2008–2011)	0.07 (2008–2010)	0.09 (2010–2011)	0.06 (2008–2011)
15	19.72 (2009–2010)	12.43 (2010–2011)	29.88 (2009–2011)	0.02 (2009–2010)	0.01 (2010–2011)	0.02 (2009–2011)
Total	402.05	236.07	342.12	0.102	0.034	0.029

Note: The "Total" row is the accumulation of the first thirteen hazards sites and does not include the last two. The years in the bracket of site No. 14 and No. 15 refers to the start and end years of variation.

Multi-Temporal Detection of Vegetation Recovery on Previously Bare Landslide Masses

In this study, vegetation recovery refers to the vegetation growth based on the previous year's hazard region. Figure 4 in Section 4.1 also illustrates the vegetation recovery in the 15 selected sites (the decreasing area from 2008 to 2011). For most sites, the recovery conditions were very positive, and vegetation began to grow on the previously bare landslide mass. Utilizing Equation (1), the vegetation recovery areas were calculated. The results are shown in Table 5. In addition, the vegetation recovery region of several sites was illustrated in Figure 6 using overlay analysis in ArcGIS platform. The fastest vegetation recovery period occurred in the first year after the earthquake, which is in line with the development of debris flows, with a recovery rate of 21.9%. Subsequently, the vegetation recovery rate trended down. In terms of the overall condition from 2008 to 2011, the average rate was 15% to 20%. The vegetation on the landslide mass may require at least five to ten years to completely recover.

Relationships between Vegetation Recovery and Debris Flow Development

The vegetation recovery rate, the debris flow development rate and the geologic hazard variation rate are shown in Figure 7. The variations of the three rates from 2008 to 2011 were in close agreement, and the vegetation recovery rate was higher than that of each site's geologic hazards variation rate. In geologic hazard areas, debris flow activity interrupts vegetation regrowth, so the faster debris flows develop, the slower the vegetation recovers, and the slower the geologic hazard varies. Figure 6 shows that, in most hazard sites, the vegetation recovery rate was higher than the debris flow development rate from 2008 to 2011, and the geologic hazard area values of the sites (except for No. 14 and 15) decreased.

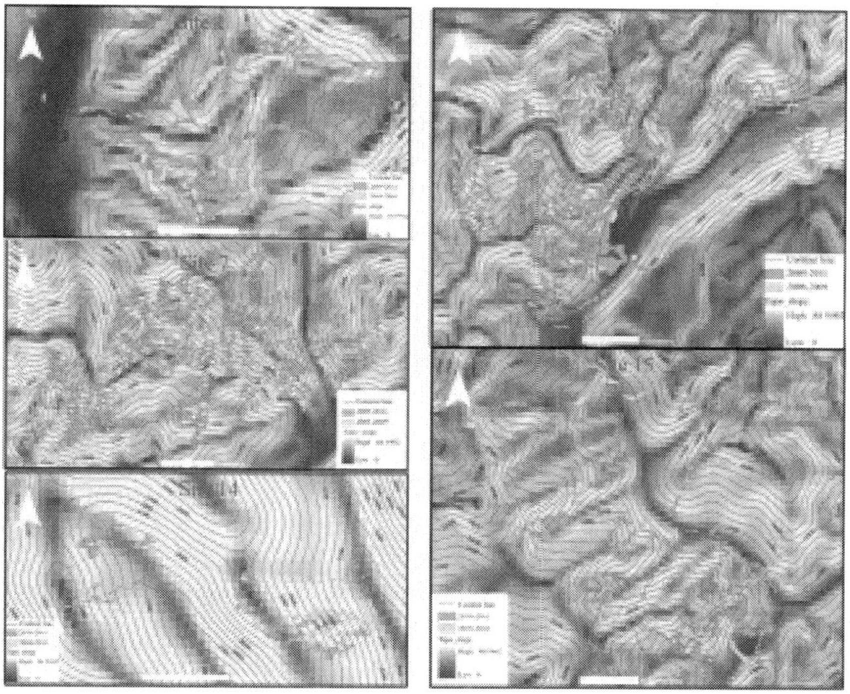

Figure 6: Vegetation recovery of site 1, 7, 10, 14. 15 from 2008 to 2009 and from 2009 to 2011.

Table 5: Area and rate of vegetation recovery on previously bare landslide masses

No.	Area of Vegetation Recovery (ha)			Rate of Vegetation Recovery (per year)		
	2008–2009	**2009–2011**	**2008–2011**	**2008–2009**	**2009–2011**	**2008–2011**
1	51.09	46.89	83.74	0.20	0.10	0.11
2	153.14	381.54	412.53	0.24	0.30	0.22
3	15.23	44.17	44.99	0.13	0.18	0.12
4	42.39	33.39	67.09	0.37	0.19	0.20
5	43.00	48.37	73.67	0.29	0.19	0.17
6	11.98	8.81	13.50	0.25	0.10	0.09
7	209.89	189.26	399.16	0.32	0.21	0.20

8	49.55	139.20	170.99	0.08	0.12	0.09
9	10.71	4.70	11.71	0.10	0.02	0.04
10	55.05	211.02	198.52	0.11	0.20	0.14
11	23.94	53.73	65.51	0.15	0.16	0.14
12	36.12	50.71	79.56	0.19	0.16	0.14
13	155.09	12.57	164.54	0.41	0.03	0.15
14	8.30 (2008–2010)	3.13 (2010–2011)	2.67 (2008–2011)	0.03 (2008–2010)	0.02 (2010–2011)	0.01 (2008–2011)
15	7.16 (2009–2010)	33.53 (2010–2011)	38.41 (2009–2011)	0.01 (2009–2010)	0.03 (2010–2011)	0.02 (2009–2011)
Total	857.18	1224.36	1785.51	0.22	0.18	0.15

Note: The "Total" row is the accumulation of the first thirteen hazards sites and does not include the last two. The years in the bracket of site No. 14 and No. 15 refers to the start and end years of variation.

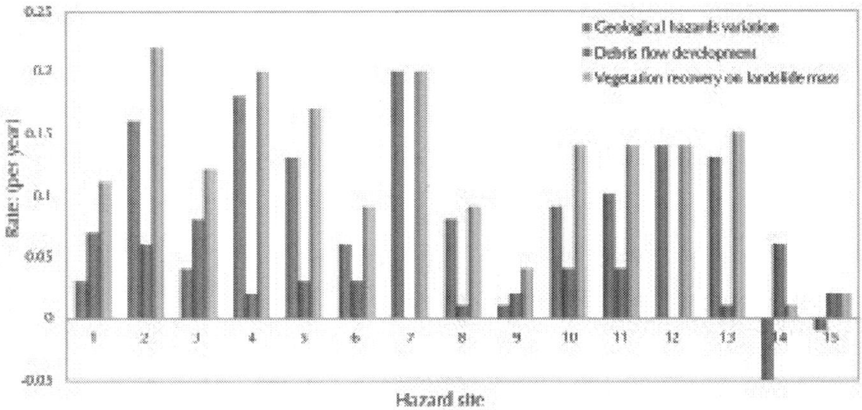

Figure 7: Comparison of hazard development rates from 2008 to 2011.

DISCUSSION

Geologic Hazards Development in the Earthquake Triggered Hazard Areas

A study by Qi *et al.* [45] showed that earthquakes (including aftershocks) and heavy rainfall were powerful causes of landslide hazards, and earthquake-induced landslides and rainfall-induced landslides could couple to form a disaster chain. A record from the Sichuan Earthquake Networks Center showed that, after the Wenchuan earthquake and up until June 2011, 88,000 aftershocks surrounding the Longmenshan fault were detected, among which more than 400 aftershocks were larger than 4.0 magnitude [46], and the occurrence of aftershocks was especially intense in the northeastern segment of the fault, which had some connection with the increased stress in this region [47]. Meanwhile, heavy rainfall occurred on 23 and 24 September 2008, reaching 250–350 mm and triggering new landslides and active old landslides, causing debris flow hazards to occur in groups [45]. Considering the four geologic hazard cases examined in Section 4.1 of this paper, the first case is relatively stable-aftershocks and heavy rainfall had little influence on the landslide mass, so the vegetation growth rate was faster than the hazard increase rate. For the second case, the geologic hazard variation is in accordance with the distribution of aftershocks and heavy rainfall. A study from Hua *et al.* [48] also showed that the aftershocks along the Longmenshan fault happened mainly within one year after the earthquake (Figure 8), and geodynamics and spatial pattern of vegetation damage was affected by rupture zones [49]. For the third and fourth cases, the geologic hazards are unstable, and the debris flow development of these two cases is consistent with human activity. For example, the anomalous variation of No. 9 and No. 14 sites was related to the Changheba Dam construction.

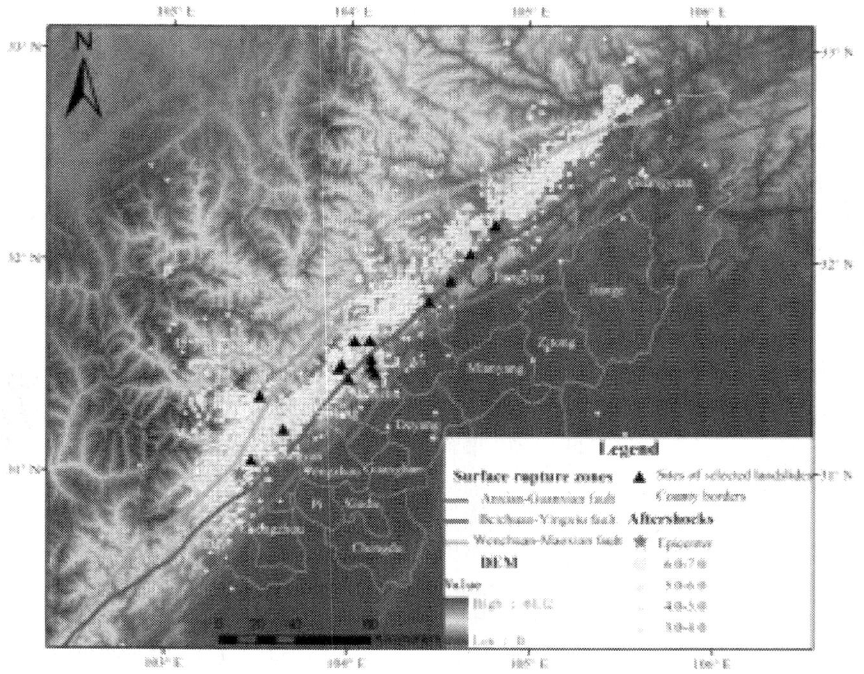

Figure 8: Map of aftershocks (up until 31 December 2008) and surface rupture zones of the Wenchuan earthquake region.

Considering the whole hazard area, the recovery of hazards was comparatively slow in the first year after the earthquake, when the area suffered the most intense aftershocks and heavy rainfalls. The rate was 11.6%, less than the average speed of recovery. After 2009, slopes became more stable and the vegetation on the landslides recovered rapidly, with an annual recovery rate of 12%. Through field investigation, we found that some artificial measures and natural vegetation recovery (Figure 9d) were important for the development of debris flows, such as the hazard prevention projects, ecological restoration projects (Figure 9a,b) and cultivated land reclamation (Figure 9c) in the earthquake affected areas slowed down the progress of debris flow in some extend.

Figure 9: (a) Hazard prevention project in Yingxiu Town, Wenchuan County (site No. 7); (b) Human activity in Wenjiagou, Qingping Town (site No. 1); (c) Reclamation on the landslide mass in Chenjiaba, Mianzhu City (site No. 4); (d) Natural vegetation recovery in Chenjiaba, Mianzhu City (site No. 4).

Debris Flow Development in the Earthquake Triggered Hazard Areas

After the Wenchuan earthquake, Chen *et al.* [44] and Cui *et al.* [30] intensively studied the characteristics of the debris flows in the Wenchuan area and found that during the first two years after the earthquake, the debris flow activity increased rapidly because of the heavy precipitation, abundant source material, and blocked valleys. Subsequently, activity of debris flows diminished as a result of the decrease in the amount of loosened soil. The authors predicted the active period would last fifteen years. Similarly, Chen *et al.* [50] studied the relationship between earthquakes and debris flows, finding that land forms, precipitation, and source material are the three key factors for debris flow initiation in the earthquake area. Based on historical observations at other earthquake areas, they predicted that the impact of the Wenchuan earthquake on local debris flows would be significant in the next five to six years and would have a lasting impact for twenty years, and that the debris flow system would eventually reach a relatively stable stage. These two conclusions effectively explain the debris flow development results in the present study. During the first year after the earthquake, the geologic hazards triggered by the earthquake, such as collapses and landslides, brought abundant loose solid masses. The surface disturbance and vegetation destruction triggered by aftershocks changed the surface water infiltration and runoff processes, and facilitated the formation of soil erosion and flooding. Meanwhile, heavy rainfall flushed the slope, carried the loose mass, and eventually formed debris flows. For example, heavy rainfalls on 24 September 2008, in Beichuan County, 13 August 2010, in Qingping Town, and 14 August 2010, in Yingxiu Town induced catastrophic debris flows. During the first year after the earthquake, debris flows occurred frequently in the earthquake-affected region with a development rate of 10.2%. After two or three years, because the loose soil source decreased, along with decreased aftershocks, the intensity of the debris flows decreased, although the debris flow activity was still in progress. The average rate of debris flow development between 2008 and 2011 was 3%.

Lithology plays a dominant role in the formation of debris flows in the study sites. The lithology data in the hazard region (Table 6) were interpreted from the China geological map and classified into three cases. The first case is glutenite and limestone, and included the No. 1, No. 2, No. 6, No. 10 and No. 11 sites. These metamorphic rocks represent good soil-forming conditions, and facilitate the formation of debris flow and vegetation recovery. The second case was for sites composed of granite and diorite with little soil development and large broken boulders after the earthquake. These conditions resulted in limited debris flow and vegetation recovery, and included the No. 3, No. 7, No. 9, No. 12 and No. 14 sites. Moreover, this lithology has a dominant proportion in the lithology of stable region. The third case was for sites composed of limestone splint rock, a sedimentary rock with moderate soil-forming conditions with less soil and small broken rock particles than glutenite. Sites No. 4, No. 5, No. 8, No. 13 and No. 15 were classified as limestone splint rock cases and had moderate debris flow and vegetation recovery.

Table 6: Lithology of the selected hazards sites

No.	Longitude	Latitude	Lithology
1	104.122	31.5519	glutenite, limestone
2	104.56	31.9247	glutenite, limestone
3	103.64	31.2122	granite, diorite
4	103.999	31.4576	limestone splint rock
5	104.146	31.4839	quartzose sandstone
6	104.811	32.1836	glutenite, limestone
7	103.464	31.0672	granite, diorite
8	104.124	31.5096	quartzose sandstone
9	103.945	31.5051	granite, diorite
10	104.441	31.8251	glutenite, limestone
11	104.669	32.052	glutenite, limestone
12	103.509	31.3697	granite, diorite
13	104.031	31.6379	limestone splint rock
14	103.959	31.5263	granite, diorite
15	104.115	31.6402	limestone splint rock

The debris flow development was also analyzed according to elevation and slope data (Figure 10). The newly developed debris flows mainly occurred in elevations below 2,000 m, accounting for 92.02% and 88.23% of the total debris flow area in 2008 to 2009 and 2009 to 2011, respectively. The debris flows also occurred on slopes of 0 to 50°, which account for 93.86% and 90.12% of the total debris flow area in 2008 to 2009 and 2009 to 2011, respectively. Large-scale landslides produced relatively large debris flows, such as the large-scale landslides in Wenjiagou, Qingping Town (site No. 1) and Yingxiu Town, Wenchuan County (site No. 7).

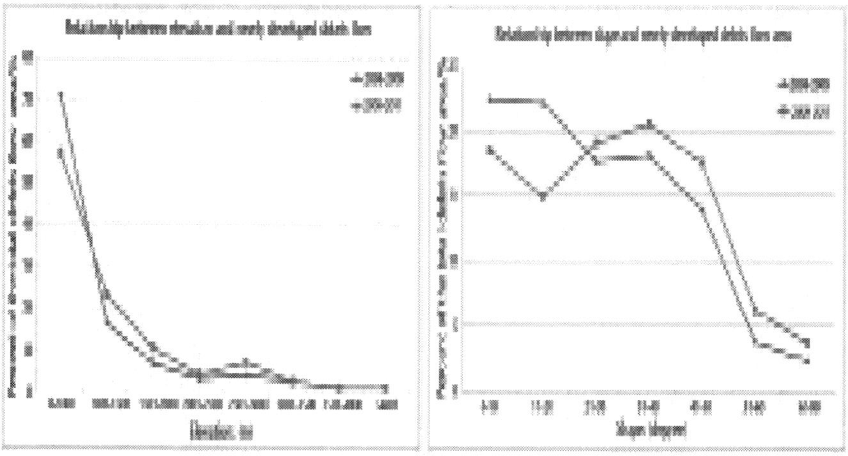

Figure 10: Relationship between newly developed debris flow area and topographic factors.

Vegetation Recovery in the Earthquake Triggered Hazard Areas

Opinions differ with regard to the potential for vegetation recovery in the hazard region. Zhang *et al.* [24] thought that an ecological restoration of the areas affected by landslides would be quite difficult because of the severe damage to the local environment caused by the earthquake. In contrast, Zhang *et al.* [51] researched the impact of the Wenchuan earthquake on biodiversity and found that despite bare rocks comprising the largest area of ground cover in the earthquake-damaged areas, a high percentage of the damaged areas contained remnant vegetation, suggesting that future vegetation recovery is likely. An experiment done by Bormann *et al.* [52] in 1981 also found that forest ecosystems had strong self-healing capabilities after the external environment had changed, as happens with earthquakes and forest fires, and the seeds buried in the soil or under the surface would play an important role in the process of vegetation recovery. During the first year after the earthquake, the seeds sprout quickly from the soil, and the dominate species are herbaceous plants. Other previous studies suggested that the presence of remnant vegetation is a driving factor for forest recovery [53,54]. Furthermore, Liu *et al.* [36] found that local vegetation productivity in heavily affected areas had recovered to 84% and 87% after one and two months, respectively.

Our results demonstrate the possibility of vegetation recovery in earthquake-affected areas even under harsh conditions in some areas, and a positive correlation was found with geological hazard variation (Figure 11). In the one to two years after the earthquake, the loose soil produced by the earthquake combined with the sliding of the entire landslide mass in some hazard regions made it easier for the vegetation to recover. Community types of shrub and grass grew quickly in this period, resulting in the fastest vegetation recovery period. Subsequently, the vegetation recovery slowed due to woody plant re-establishment. The recovery patterns in the hazard areas were complex, with factors such as the terrain and soil also affecting the recovery's progress [3] and making vegetation recovery difficult (as with the No. 14 and No. 15 sites in Table 5).

Similarly, the vegetation recovery was analyzed according to elevation and slope data (Figure 12). The vegetation recovery mainly occurred in elevations below 3,000 m, accounting for 99.27% and 99.15% of the total debris flow area in 2008 to 2009 and 2009 to 2011, respectively, and the peak occurred between the elevation of 1001–1500 m because of the abundant accumulational soil mass, which was suitable for vegetation growth, produced by the landslides and debris flows. The slope of the surface also have a obvious influence on the progress of vegetation recovery, and more than 60% of the vegetation recovery were concentrated on slopes between 31° and 50°, the peak occurred between the slope of 41–50°, Lin *et al.* [55] and Lu *et al.* [3] also observed similar results, they suggested that grass species that survived during this slope range facilitated the vegetation growth when adequate rainfall was supplied to the area. In addition, the good soil holding and storing water condition for the plants in the hazard sites is also critical to the vegetation recovery.

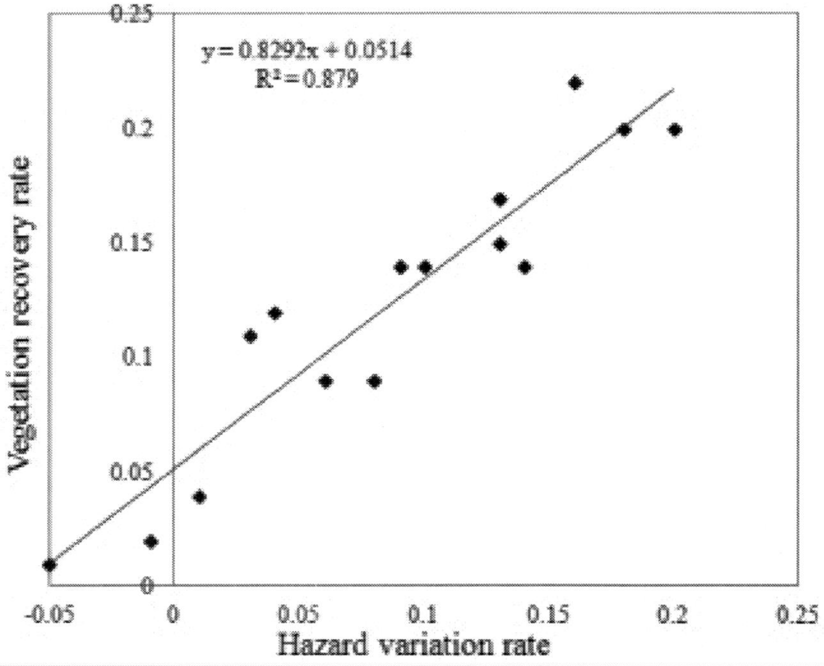

Figure 11: Relationship between vegetation recovery and geological hazards development from 2008 to 2011.

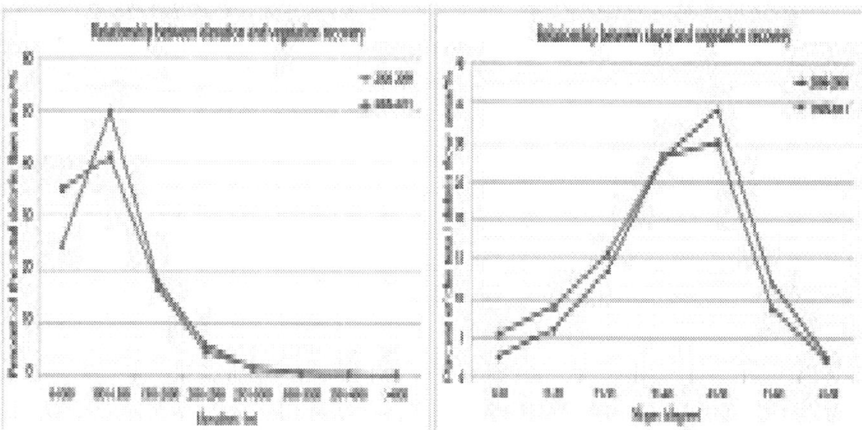

Figure 12: Relationship between vegetation recovery and topographic factors.

CONCLUSIONS

Remote sensing technology can improve the efficiency of disaster monitoring. This study used airborne optical images to monitor the development of secondary hazards and vegetation recovery in post-earthquake areas, and the results were validated through field investigation.

During the first year after the earthquake, debris flow activity occurred frequently with high intensity. After the source material became less available and the slope structure stabilized, the intensity of the debris flows trended down. The average annual rate of newly developed debris flows between 2008 and 2011 was approximately 3%. After the first year, the debris flow and vegetation recovery activity gradually diminished, and the hazard area gradually decreased to 12%. The lithology and topography had an influence on the debris flow development process.

Observations in the post-earthquake hazard areas from 2008 to 2011 showed vegetation recovery in these areas was possible, and the greatest vegetation recovery occurred two or three years after the earthquake. However, the vegetation recovery rate was related to debris flow development, and was in agreement with post-earthquake geologic hazard development. The fastest vegetation recovery period was the first one to two years after the earthquake. Afterwards, the vegetation recovery gradually slowed. The average recovery rate between 2008 and 2011 was 15% to 20%. Thus, the results of this paper predict that at least five to ten years are needed to achieve total vegetation recovery. In addition, the elevation below 1500 m and slope between 31° and 50° has the fastest rate of vegetation growth.

Given the various performances of the different hazard sites, future studies should consider the whole disaster area as the research area. Other factors such as the soil structure [56,57], terrain [58,59], and regional climate [60,61] provide more accurate data for decision making. To acquire a better understanding of the hazard development and vegetation restoration processes, longer-term observations, such as ten to twenty years, must be conducted [62].

ACKNOWLEDGMENTS

The authors gratefully acknowledge financial support provided for this research from the Interdisciplinary and Collaborative S&T Innovation Research Team on Advance Earth Observation System CAS and the External Cooperation Program of the Chinese Academy of Sciences (GJH21123). We also thank all the anonymous reviewers for their valuable comments, which substantially improved our paper.

AUTHOR CONTRIBUTIONS

Zhenwang Li and Quanjun Jiao performed the analyses and prepared the paper. Liangyun Liu provided the technical guidance and contributed to the discussion. Zhenwang Li, Huan Tang, and Quanjun Jiao helped the collection of field data. Tong Liu helped provide airborne remote sensing images.

REFERENCES

1. Cui, P.; Zhu, Y.-Y.; Han, Y.-S.; Chen, X.-Q.; Zhuang, J.-Q. The 12 May Wenchuan earthquake-induced landslide lakes: Distribution and preliminary risk evaluation. *Landslides* 2009, *6*, 209–223.
2. Han, J.; Wu, S.; He, S.; Sun, W.; Zhang, C.; Wang, T.; Yang, J.; Si, J. Basal characteristics and formation mechanisms of geological hazards triggered by the May 12, 2008 Wenchuan earthquake with a moment magnitude of 8.0. *Earth Sci. Fontiers* 2009, *16*, 306–326.
3. Lu, T.; Zeng, H.; Luo, Y.; Wang, Q.; Shi, F.; Sun, G.; Wu, Y.; Wu, N. Monitoring vegetation recovery after China's May 2008 Wenchuan earthquake using landsat tm time-series data: A case study in Mao County. *Ecol. Res.* 2012, *27*, 955–966.
4. Cui, P.; Wei, F.; Chen, X.; He, S. Geo-hazards in Wenchuan earthquake area and countermeasures for disaster reduction. *Bull. Chin. Acad. Sci.* 2008, *23*, 317–323.
5. Li, Z. The state of the art of the research on seismic landslide hazard at home and abroad. *J. Catastrophol.* 2003, *18*, 64–70. [Google Scholar]

6. Tronin, A. Satellite remote sensing in seismology. A review. *Remote Sens.* 2009, *2*, 124–150.

7. Dong, Y.; Li, Q.; Dou, A.; Wang, X. Extracting damages caused by the 2008 Ms 8.0 Wenchuan earthquake from SAR remote sensing data. *J. Asian Earth Sci.* 2011, *40*, 907–914.

8. Ehrlich, D.; Guo, H.D.; Molch, K.; Ma, J.W.; Pesaresi, M. Identifying damage caused by the 2008 Wenchuan earthquake from VHR remote sensing data. *Int. J. Digit. Earth* 2009, *2*, 309–326.

9. Ellen, R.; Robert, K.; Kyu-Seok, W. Remote sensing observations of landslides and ground deformation from the 2004 Niigata Ken Chuetsu earthquake. *Jpn. Geotech. Soc.* 2006, *46*, 831–842.

10. Fu, B.; Ninomiya, Y.; Lei, X.; Toda, S.; Awata, Y. Mapping active fault associated with the 2003 Mw 6.6 Bam (Se Iran) earthquake with ASTER 3D images. *Remote Sens. Environ.* 2004, *92*, 153–157.

11. Guo, H.; Liu, L.; Fan, X.; Li, X. Study of earth observation for disaster reduction in Wenchuan and Yushu earthquakes. *Geol. J. China Univ.* 2011, *17*, 1–12.

12. Yusuf, Y.; Matsuoka, M.; Yamazaki, F. Damage assessment after 2001 Gujarat earthquake using Landsat-7 satellite images. *J. Indian Soc. Remote Sens.* 2001, *29*, 17–22.

13. Safari, H.O.; Pirasteh, S.; Pradhan, B.; Gharibvand, L.K. Use of remote sensing data and GIS tools for seismic hazard assessment for shallow oilfields and its impact on the settlements at Masjed-i-Soleiman area, Zagros mountains, Iran. *Remote Sens.* 2010, *2*, 1364–1377.

14. Liou, Y.A.; Sha, H.C.; Chen, T.M.; Wang, T.S.; Li, Y.T.; Lai, Y.C.; Chiang, M.H.; Lu, L.T. Assessment of disaster losses in rice paddy field and yield after Tsunami induced by the 2011 great east Japan earthquake. *J. Mar. Sci. Technol.-Taiwan* 2012, *20*, 618–623.

15. Kulawardhana, R.W. Remote sensing and GIS technologies for monitoring and prediction of disasters. *Int. J. Digit. Earth* 2011, *5*, 88–90.

16. Gorum, T.; Fan, X.; van Westen, C.J.; Huang, R.Q.; Xu, Q.; Tang, C.; Wang, G. Distribution pattern of earthquake-induced landslides triggered by the 12 May 2008 Wenchuan earthquake. *Geomorphology* 2011, *133*, 152–167.

17. Zhang, B.; Jiao, Q.; Wu, Y.; Zhang, W. Estimating soil erosion changes in the Wenchuan earthquake disaster area using geo-spatial information technology. *J. Appl. Remote Sens.* 2009, *3*.
18. Klose, C.D. Evidence for anthropogenic surface loading as trigger mechanism of the 2008 Wenchuan earthquake. *Environ. Earth Sci.* 2011, *66*, 1439–1447.
19. Li, Z.; Chen, Q.; Zhou, J.; Tian, B. Analysis of synthetic aperture radar image characteristics for seismic disasters in the Wenchuan earthquake. *J. Appl. Remote Sens.* 2009, *3*.
20. Tong, X.; Sandwell, D.T.; Fialko, Y. Coseismic slip model of the 2008 Wenchuan earthquake derived from joint inversion of interferometric synthetic aperture radar, GPS, and field data. *J. Geophys. Res.: Solid Earth* 2010, *115*.
21. Lei, L.; Liu, L.; Zhang, L.; Bi, J.; Wu, Y.; Jiao, Q.; Zhang, W. Assessment and analysis of collapsing houses by aerial images in the Wenchuan earthquake. *J. Remote Sens* 2010, *14*, 333–344.
22. Liu, L.; Wu, Y.; Zuo, Z.; Chen, Z.; Wang, X.; Zhang, W. Monitoring and assessment of barrier lakes formed after the Wenchuan earthquake based on multitemporal remote sensing data. *J. Appl. Remote Sens.* 2009, *3*.
23. Sato, H.; Harp, E. Interpretation of earthquake-induced landslides triggered by the 12 May 2008, M7.9 Wenchuan earthquake in the beichuan area, sichuan province, China using satellite imagery and google earth. *Landslides* 2009, *6*, 153–159.
24. Zhang, W.; Lin, J.; Peng, J.; Lu, Q. Estimating Wenchuan earthquake induced landslides based on remote sensing. *Int. J. Remote Sens.* 2010, *31*, 3495–3508.
25. Wang, X.; Dou, A.; Ding, X. Study on Quantitative Earthquake Damage of Dujiangyan City, Caused by 2008 MS = 8.0 Wenchuan, China Earthquake Based on Aerial Imagery. In Proceedings of the 2010 IEEE International Geoscience and Remote Sensing Symposium (IGARSS), Honolulu, HI, USA, 25–30 July 2010; pp. 2743–2746.
26. Brown, D.; Saito, K.; Liu, M.; Spence, R.; So, E.; Ramage, M. The use of remotely sensed data and ground survey tools to assess damage and monitor early recovery following the 12.5.2008 Wenchuan earthquake in China. *Bull. Earthq. Eng.* 2011, *10*, 741–764.

27. Cao, C.; Chang, C.; Xu, M.; Zhao, J.; Gao, M.; Zhang, H.; Guo, J.; Guo, J.; Dong, L.; He, Q.; *et al*. Epidemic risk analysis after the Wenchuan earthquake using remote sensing. *Int. J. Remote Sens.* 2010, *31*, 3631–3642.

28. Liou, Y.-A.; Kar, S.K.; Chang, L. Use of high-resolution formosat-2 satellite images for post-earthquake disaster assessment: A study following the 12 May 2008 Wenchuan earthquake. *Int. J. Remote Sens.* 2010, *31*, 3355–3368.

29. Dong, J.-J.; Liou, Y.-A.; Chang, L.-Y.; Lee, C.-T.; Liao, J.-J.; Pa, Y.-W. Application of satellite images and DEM for the hazard assessment of landslide dams. *J. Photogramm. Remote Sens.* 2010, *15*, 3–15.

30. Cui, P.; Zhuang, J.; Chen, X.; Zhang, J.; Zhou, X. Characteristics and countermeasures of debris flow in Wenchuan area after the earthquake. *J. Sichuan Univ. (Eng. Sci. Ed.)* 2010, *42*, 10–19. (In Chinese).

31. Li, Y.; Huang, R.; Densmore, A.L.; Zhou, C.; Cao, S. Basic features and research progresses of Wenchuan MS 8.0 earthquake *J. Sichuan Univ. (Eng. Sci. Ed.)* 2009, *41*, 7–25.

32. Yang, C.; Ren, X.; Huang, H. The vegetation damage assessment of the Wenchuan earthquake of May 2008 using remote sensing and GIS. *Nat. Hazards* 2011, *62*, 45–55.

33. Chigira, M.; Wu, X.; Inokuchi, T.; Wang, G. Landslides induced by the 2008 Wenchuan earthquake, Sichuan, China. *Geomorphology* 2010, *118*, 225–238.

34. Huang, R.; Li, W. Analysis of the geo-hazards triggered by the 12 May 2008 Wenchuan earthquake, China. *Bull. Eng. Geol. Environ.* 2009, *68*, 363–371.

35. Ge, Y.; Xu, J.; Liu, Q.; Yao, Y.; Wang, R. Image interpretation and statistical analysis of vegetation damage caused by the Wenchuan earthquake and related secondary disasters. *J. Appl. Remote Sens.* 2009, *3*.

36. Liu, Y.; Liu, R.; Ge, Q. Evaluating the vegetation destruction and recovery of Wenchuan earthquake using MODIS data. *Nat. Hazards* 2010, *54*, 851–862.

37. Xu, J.; Lu, Y. Meta-synthesis pattern of post-disaster recovery and reconstruction: Based on actual investigation on 2008 Wenchuan earthquake. *Nat. Hazards* 2011, *60*, 199–222.

38. Guo, H.; Liu, L.; Lei, L.; Wu, Y.; Li, L.; Zhang, B.; Zuo, Z.; Li, Z. Dynamic analysis of the Wenchuan earthquake disaster and reconstruction with 3-year remote sensing data. *Int. J. Digit. Earth* 2010, *3*, 355–364.

39. Cruden, D.M. A simple definition of a landslide. *Bull. Int. Assoc. Eng. Geol.* 1991, *43*, 27–29.

40. Turner, A.K.; Schuster, R.L. *Landslides: Investigation and Mitigation*; Transportation Research Board, Special Report 247. National Academy Press: Washington, DC, USA, 1996; pp. 129–177.

41. Highland, L.M.; Bobrowsky, P. *The Landslide Handbook: A Guide to Understanding Landslides*; U.S. Geological Survey Circular 1325: Reston, VA, USA, 2008; p. 129.

42. Highland, L.; Ellen, S.D.; Christian, S.B.; Brown, W.M., III. *Debris-Flow Hazards in the United States*; U.S. Department of the Interior, U.S. Geological Survey: Denver, CO, USA, 1997.

43. Ren, Z.; Lin, A. Co-seismic landslides induced by the 2008 Wenchuan magnitude 8.0 earthquake, as revealed by ALOS PRISM and AVNIR2 imagery data. *Int. J. Remote Sens.* 2010, *31*, 3479–3493.

44. Chen, X.; Cui, P.; Li, Y.; Gao, Q.; Zhao, W. Mountain hazard induced by Wenchuan earthquake and its long-term development trends of Ganxi Gully, Beichuan. *J. Sichuan Univ. (Eng. Sci. Ed.)* 2010, *42*, 22–32. (In Chinese).

45. Qi, X.; Tang, C.; Chen, Z.; Shao, C. Coupling analysis of control factors between earthquake-induced landslides and subsequent rainfall-induced landslides in epicenter area of Wenchuan earthquake. *J. Eng. Geol.* 2012, *20*, 522–531.

46. Yi, G.; Long, F.; Zhang, Z. Spatial and temporal variation of focal mechanisms for aftershocks of the 2008 M_s8.0 Wenchuan earthquake. *Chin. J. Geophys.* 2012, *55*, 1213–1227. (In Chinese).

47. Li, Y.; Chen, L.; Lu, Y.; Zhan, Z. Numerical simulation on influences of Wenchuan earthquake on the stability of faults in the neighborhood. *Earth Sci. (J. China Univ. Geosci.)* 2013, *38*, 398–410. (In Chinese).

48. Hua, W.; Chen, Z.; Li, Z.; Zhao, C.; Wang, Q.-C. Seismic triggering and the aftershock distribution of the Wenchuan M8.0 earthquake. *Earthquake* 2009, *29*, 33–39.

49. Masoud, A.; Koike, K. Relationship between remotely sensed vegetation change and fracture zones induced by the 2008 Wenchuan earthquake, China. *J. Earth Sci.* 2013, *24*, 282–296.

50. Chen, N.-S.; Hu, G.-S.; Deng, M.-F.; Zhou, W.; Yang, C.-L.; Han, D.; Deng, J.-H. Impact of earthquake on debris flows—A case study on the Wenchuan earthquake. *J. Earthq. Tsunami* 2011, *5*, 493–508.

51. Zhang, J.; Hull, V.; Xu, W.; Liu, J.; Ouyang, Z.; Huang, J.; Wang, X.; Li, R. Impact of the 2008 Wenchuan earthquake on biodiversity and giant panda habitat in Wolong Nature Reserve, China. *Ecol. Res.* 2011, *26*, 523–531.

52. Bormann, F.H.; Likens, G.E. *Pattern and Process in a Forested Ecosystem: Disturbance, Development, and the Steady State Based on the Hubbard Brook Ecosystem Study*; Springer-Verlag: New York, USA, 1994; p. 207.

53. Ferguson, B.G.; Vandermeer, J.; Morales, H.; Griffith, D.M. Post-agricultural succession in el petén, guatemala. *Conserv. Biol.* 2003, *17*, 818–828.

54. Holl, K.D.; Loik, M.E.; Lin, E.H.V.; Samuels, I.A. Tropical montane forest restoration in Costa Rica: Overcoming barriers to dispersal and establishment. *Restor. Ecol.* 2000, *8*, 339–349.

55. Lin, W.-T.; Chou, W.-C.; Lin, C.-Y.; Huang, P.-H.; Tsai, J.-S. Vegetation recovery monitoring and assessment at landslides caused by earthquake in central Taiwan. *For. Ecol. Manag.* 2005, *210*, 55–66.

56. Mhaske, S.Y.; Choudhury, D. GIS-based soil liquefaction susceptibility map of Mumbai city for earthquake events. *J. Appl. Geophys.* 2010, *70*, 216–225.

57. Brocca, L.; Ponziani, F.; Moramarco, T.; Melone, F.; Berni, N.; Wagner, W. Improving landslide forecasting using ASCAT-derived soil moisture data: A case study of the Torgiovannetto landslide in central Italy. *Remote Sens.* 2012, *4*, 1232–1244.

58. Kamp, U.; Growley, B.J.; Khattak, G.A.; Owen, L.A. GIS-based landslide susceptibility mapping for the 2005 Kashmir earthquake region. *Geomorphology* 2008, *101*, 631–642.

59. Lee, S.; Talib, J. Probabilistic landslide susceptibility and factor effect analysis. *Environ. Geol.* 2005, *47*, 982–990.

60. Wang, L.; Shum, C.K.; Simons, F.J.; Tassara, A.; Erkan, K.; Jekeli, C.; Braun, A.; Kuo, C.; Lee, H.; Yuan, D.-N. Coseismic slip of the 2010 Mw 8.8 Great Maule, Chile, earthquake quantified by the inversion of GRACE observations. *Earth Planet. Sci. Lett.* 2012, *335–336*, 167–179.
61. Li, A.; Wang, A.; Liang, S.; Zhou, W. Eco-environmental vulnerability evaluation in mountainous region using remote sensing and GIS—A case study in the upper reaches of Minjiang river, China. *Ecol. Model.* 2006, *192*, 175–187.
62. Lu, T.; Shi, F.; Sun, G.; Luo, Y.; Wang, Q.; Wu, Y.; Wu, N. Reconstruction of the Wenchuan earthquake-damaged ecosystems: Four important questions. *Chin. J. Appl. Environ. Biol.* 2010, *16*, 301–304.

Chapter 8

Epidemiologic Methods Lessons Learned from Environmental Public Health Disasters: Chernobyl, the World Trade Center, Bhopal, and Graniteville, South Carolina

Erik R. Svendsen[1, 2], Jennifer R. Runkle[2, 3], Venkata Ramana Dhara[4], Shao Lin[5], Marina Naboka[6], Timothy A. Mousseau[7] and Charles Bennett[8]

[1]Department of Global Environmental Health Sciences, Tulane University School of Public Health and Tropical Medicine, Tulane University Health Sciences Center, ENHS SL-29, 1440 Canal St., Suite 2100, New Orleans, LA 70112, USA

[2]Department of Epidemiology and Biostatistics, University of South Carolina, Columbia, SC 29208, USA

[3]Nell Hodgson Woodruff School of Nursing, Emory University, Atlanta, GA 62234, USA

[4]Department of Environmental Health, Emory University, Atlanta, GA 62234, USA

[5]Bureau of Occupational and Environmental Epidemiology, State University of New York-Albany, Albany, NY 84870, USA

[6]Radioecological Center of the National Academy of Sciences, Kiev 04050, Ukraine

[7]Department of Biological Sciences, University of South Carolina, Columbia, SC 29208, USA

[8]College of Pharmacy, University of South Carolina, Columbia, SC 29208, USA

ABSTRACT

Background

Environmental public health disasters involving hazardous contaminants may have devastating effects. While much is known about their immediate devastation, far less is known about long-term impacts of these disasters. Extensive latent and chronic long-term public health effects may occur. Careful evaluation of contaminant exposures and long-term health outcomes within the constraints imposed by limited financial resources is essential. *Methods*: Here, we review epidemiologic methods lessons learned from conducting long-term evaluations of four environmental public health disasters involving hazardous contaminants at Chernobyl, the World Trade Center, Bhopal, and Graniteville (South Carolina, USA). *Findings*: We found several lessons learned which have direct implications for the on-going disaster recovery work following the Fukushima radiation disaster or for future disasters. Interpretation: These lessons should prove useful in understanding and mitigating latent health effects that may result from the nuclear reactor accident in Japan or future environmental public health disasters.

INTRODUCTION AND METHODS

Despite our preparedness efforts, disasters are unpredictable, unannounced and unexpected. Occasionally disasters result in the release of a toxicological hazard into the affected environment. Such events may be considered environmental public health disasters (EPHD). Disasters are chronicled by stories of acute devastation and initial large-scale humanitarian relief efforts. Shortly thereafter, attention wanes and humanitarian efforts dwindle, while adverse long-term health effects may follow. Such effects are frequently poorly understood and even less so concerning long-term repeated exposures to a contaminated environment resulting from a disaster. Epidemiologic studies of EPHDs resulting from accidental or intentional releases of hazardous contaminants provide important information on long-term toxic health effects [1,2,3,4], though methodological challenges such as mitigating selection biases, exposure assessment, and community mistrust present important obstacles to comprehensively evaluating these long-term health effects. The purpose of this paper is to share some practical lessons learned during the public health recovery work after four EPHDs (Table 1).

Table 1: Epidemiologic methods lessons learned from four Environmental Public Health Disasters

Disaster	Location	Year	Population Scale	Exposure	Outcomes	Lessons Learned
Bhopal	India	1984	Hundreds of Thousands	Methyl isocyanate	Multiple systems	Sometimes simple studies are sufficient.
Chernobyl	Ukraine	1986	Hundreds of Thousands	Radioactive fallout	Multiple systems	Sometimes public health interventions can facilitate longitudinal studies. Sometimes "ecological epidemiology" can be a useful alternative to human studies.

World Trade Center Collapse	New York City	2001	Thousands	Caustic dusts	Mostly pulmonary	Sometimes careful inclusion of an unaffected control population is better than comparing distance to exposure only.
Graniteville	South Carolina	2005	Many hundreds	Chlorine gas (acidic)	Mostly pulmonary	Sometimes using longitudinal occupational health cohorts can augment the population-based studies

It is our hope that these practical lessons will contribute to both current disaster recovery work, like with the Fukushima disaster, and the public health recovery work after future unfortunate disasters. We reviewed epidemiologic lessons learned from evaluations of four EPHDs in Europe, North America, and Asia which involved hazardous contaminants. Investigators directly involved with developing and implementing epidemiological studies following EPHDs in Bhopal, Chernobyl, New York City, and South Carolina collaboratively reviewed both their experiences and those of their collaborators, and discussed them collectively. The disasters in Bhopal and South Carolina illustrate impacts from exposure to a toxic gas while the World Trade Center (WTC) disaster in New York City demonstrates the effects of exposure to caustic dust particulates. The Chernobyl disaster illustrates impacts of a radiological accident involving emissions of aerosols and gasses with both acute and chronic impacts on surrounding regions. These lessons learned are not comprehensive accounts of the public health and epidemiological activities after these disasters. Rather, they are the accounts of scientists intricately involved in the research and recovery process with general discussions of the relevant issues illustrated within.

RESULTS

Bhopal

Lesson—sometimes simple studies are sufficient. In 1984 an industrial accident involving a 27 metric ton gas leak of methyl isocyanate (MIC) occurred at a pesticide plant in Bhopal, India [5]. This event affected over 500,000 people in nearby residential areas [6, 7]. The magnitude of the accident, political and governmental inexperience in dealing with such a large-scale event, and the rapid dissolution of medical and social infrastructure impeded immediate evaluation of the toxic effects [8]. A decade later a transnational collaboration, the International Medical Commission on Bhopal (IMCB), was established to perform a cross-sectional epidemiologic study of the health effects of MIC exposure. To avoid self-selection bias, we performed simple random sampling for participant selection. Plume dispersion models (computer models using the available source, weather and topography data to predict where a chemical plume is dispersed within the atmosphere) were very imprecise due to the lack of sufficient local weather data. Therefore, exposure estimation was retrospectively assessed using three exposure indices: (1) physical activity during the event (e.g., fleeing gas *vs.* sheltering), (2) exposure duration, and (3) distance of residence from the pesticide plant. Current and past pulmonary symptoms were queried and an objective outcome assessment was performed: simple spirometry [7]. Hence, this study was susceptible to recall biases. Barriers to health care, traumatic stress, and disappointment with official policies were epidemic at the time of the study. It is commonly assumed that these issues may lead to inflated symptom reporting post-disaster. This study observed a decrease in both symptoms and objective spirometry deficits associated with increasing distance from the accident, suggesting that exposure misclassification from symptom over-reporting was likely non-differential [7] and not suggestive of recall bias. Though objective exposure measures were unobtainable the IMCB successfully executed a crude exposure assessment which suggested an association between MIC exposure and latent respiratory and pulmonary health effects [7]. The study showed that simple measures may be sufficient to link individual exposure estimates to observed health outcomes [6]. In most cases, a longitudinal study is the preferred method for assessing exposure and disease outcome among disaster victims over time. However, for Bhopal, the cross-sectional study circumvented issues of time, money, and changes in secular trends which can plague longitudinal cohort studies. Furthermore, a cross-sectional study can easily transition into a "cross-sectional cohort study" [9]; in which, a source population is sampled cross-sectionally and followed by retrospective assessments of participants" exposure and outcome histories over time. This type of study is ideal for chronic and recurrent diseases with low death rates, often associated with technological disasters [9]. For latent analysis of exposure-response, an individual exposure index based on duration of exposure, location, and physical activity is a relatively easy strategy to implement and can be used to validate the relationship between individual exposure and subjective and objective measures.

Chernobyl

Lesson—sometimes public health interventions can facilitate longitudinal studies. Soon after the Bhopal tragedy, in 1986 a nuclear reactor accident releasing radioactive isotopes including ^{131}I, ^{137}Cs, ^{90}Sr and ^{239}Pu occurred at the Chernobyl Nuclear Power Plant in the Ukraine. The accident was declared the worst nuclear power disaster in history, although the full impact of the Fukushima disaster remains unknown. Global plume dispersion models and aerial gamma-spectroscopy revealed high radioactive fallout in the Western Soviet Union, Eastern Europe, Western Europe, Northern Europe, and even North America, significantly contaminating more than 200,000 km^2. Although impacts from the transient gaseous emissions of ^{131}I on thyroid cancer rates have been well-documented [10,11,12], radiation exposure stemming from aerosolized radioactive heavy metals which is persistent in the soil due to long half-lives (30 and 29 years for ^{137}Cs and ^{90}Sr, respectively) has been less studied. Continuing transformation of plutonium to highly radioactive Am isotopes will persist for many years [13].

Early epidemiologic assessment of the Chernobyl disaster faced significant barriers. These included the impact of radionuclides moving through the food chain, lack of data among individuals who moved out of contaminated areas, variability in population-levels of radiosensitivity, missing data on the impact of individual radionuclides and their subsequent synergistic impact with other environmental hazards, and uncertainty in the consequences of nominal ionizing radiation doses and internal radiation exposures [14]. Methodological limitations included poor exposure reconstruction, inherent fluctuations in physical and chemical properties of radionculides, and inability to capture external and internal radiation levels of exposed populations. Furthermore, data collection failed to follow standardized scientific protocols in the rush to collect data within the immediate disaster period, resulting in questionable reliability.

Soviet censorship and constraints related to historical data acquisition forced Ukrainian researchers to develop novel and cost-effective approaches to conducting epidemiologic assessments of Chernobyl. Bypassing the expense and time it would take to assemble a cohort, obtain detailed clinical and laboratory information, and accrue enough follow-up time to detect exposure-response associations, one approach harvested pediatric health data from existing Ukrainian public health surveillance systems. In 1986 the Soviets mandated that children age 0–14 years residing in the most contaminated and unevacuated areas of Russia, Belarussia, and Ukraine receive annual medical examinations. These were maintained by the Ukrainian government after the fall of the Soviet Union until 2006 when the funding for public health program was cut and the mandate halted. As a result, the Ukrainian Ministry of Health collected twenty years of longitudinal data on medical history, physical exams, and laboratory analyses for over 1 million children from the most contaminated regions of Ukraine. These data were collected for public health purposes with no initial plans to use them for scientific research. Transcription of only six years of such data from only one affected region has generated novel scientific findings when linked with exposure data and has been recently transitioned into an elective longitudinal cohort study [15,16,17,18]. Much more could be learned from these health records and the longitudinal cohort study that was born from them.

Due to dependence on subsistence farming in the most contaminated areas and the poor compliance with public health recommendations of avoiding local food sources, potential ionizing radiation exposure through ingestion of local food was a major public health concern. Therefore, public health officials tested the use of local soils to estimate personal radiation exposure. Soil measurements and the spatial distribution of radiation data were collected over several years (1987–1993). The Chernobyl Sasakawa Health and Medical Cooperation project provided evidence that individual ^{137}Cs levels were highly correlated with residential soil contamination levels [19]. However, such soil exposure information was not readily available for research purposes until relatively recently.

To glean research from the available public health records, personal exposure estimates derived from residential soil contamination was linked to the public health screening data from the 6 years of abstracted data previously mentioned. This research used six-years (1993–1998) of Ukrainian Ministry of Health surveillance data representing ~75% of the children within one region (although the health screenings were mandated there was no enforcement). Many children born both before and after the 1986 accident were included in the cohort, allowing researchers to distinguish between environmental and in utero exposures. Results revealed that poor health outcomes among exposed children were associated with high levels of local soil ^{137}Cs [16, 20]. In response to these findings, public health officials quickly communicated the risks of local food consumption to at-risk regions yet again.

Lesson—sometimes "ecological epidemiology" can be a useful alternative to human studies. Because people are not the only organisms exposed, a second approach to evaluating the health impact of the Chernobyl disaster involved long-term ecological assessments of many sentinel species in areas with varying levels of radioactive contamination [21]. Groups used include birds, butterflies, dragonflies, grasshoppers, spiders, amphibians, reptiles, mammals and plants [22,23,24,25]. Such "ecological epidemiology" can be used to assess animal sentinels, or bioindicators of exposure, as estimates of potential human health effects and risks within the disaster region. Sensitivity analyses suggest that birds are particularly sensitive and easily measurable sentinels for the ecological effects of radionuclides in the environment [26] and recent comparisons with impacts on animals in Fukushima suggest that birds and butterflies are impacted both by acute radiotoxicity and chronic mutation accumulation effects [22,27]. These studies are not impacted by the traumatic stress effects of surviving a disaster or living within the context of disaster, nor how the resulting health behaviors may modify toxicological effects. Such potential counfounders require careful assessment in human disaster studies [28]. These sentinel animal studies are relatively cost-effective, out-migration is rare if studying a species with a small home range, and clinical, laboratory, and pathology studies can be conducted efficiently. As in the prior human epidemiology study [16], radiation exposure can be proxied by measurements of ionizing radiation in the soil. Now, over a quarter century after the accident, sentinel animal studies have indicated effects of ionizing radiation on brain size [29], tumors and other developmental abnormalities [30], decreased sperm motility and higher frequencies of defective sperm [31], and higher bacterial loads on birds who resided closer to the disaster site. Such studies provide empirical evidence that ionizing radiation likely has long-term sentinel health effects, not simply effects due to poor health behaviors or psychological stress [32].

The World Trade Center

Lesson—sometimes careful inclusion of an unaffected control population is better than comparing distance to exposure only. Many studies now document that exposure to extremely caustic dust from the collapse of the World Trade Center (WTC) has increased the long-term medical needs of victims, responders, and clean-up workers [33,34,35,36,37,38,39]. One set of studies evaluated changes in respiratory health from excessive dust exposure among residents living near the former WTC in New York City [35,40]. Methodological challenges to this assessment included identification of exposure-associated outcomes, exposure misclassification, recall and selection bias. Similar to the Bhopal disaster, direct exposure measurements were unavailable; therefore residential distance from the WTC was used as an exposure proxy. However, unlike the Bhopal disaster, the target population included people both affected and unaffected (control): 9,200 residences within 1.5 kilometers of the former WTC (the affected area) and 1,000 residences in upper Manhattan more than 9 kilometers from the site (control area). To minimize selection bias and improve response rates, participation was encouraged among individuals both with and without breathing problems. By comparing demographics between respondents and the overall population, exclusion criteria were instituted to minimize differential selection by study area. Recruitment was intensified in "target" areas to improve response rates and better evaluate participation bias. Response rates were similar in both the affected and control areas. Reporting bias was minimized by qualitatively and quantitatively asking symptomology questions incorporating questions on exact time periods, severity, and exposure frequency [40]. The expectation of exposure misclassification was attenuated by excluding participants with evidence of residential mobility and other exposures not related to their geographic location [40]. Area analyses showed higher elevated rates of respiratory diseases in both recruitment targeted and non-targeted areas directly impacted, reducing concerns about selection bias. Assessment of distance from the event successfully informed public health officials of the disaster's health effects on neighboring communities.

Lesson—sometimes using longitudinal occupational health cohorts can augment the population-based studies. A second set of studies evaluated long-term consequences of exposure to WTC dust by characterizing prospective trends in pulmonary function for seven years following the disaster [41]. This study examined changes in forced expiratory volume in 1 second (FEV1) among Federal Department of New York (FDNY) rescue workers. This study used occupational health surveillance data, including before and after spirometric measurement, on 92% of the 13,954 FDNY personnel (11,868 firefighters and 2,086 EMS workers) present at the WTC site between 11 September and 24 September, 2001. A robust natural experiment design which contrasted before and after trends in lung function (*i.e.*, interrupted longitudinal cohort) determined little or no recovery of average lung function decline during the 6-year follow-up period. From 2002 through 2008, FEV1 values continued to decline. The overall loss in lung function from early 2001 until late 2008 averaged almost 600 mL for firefighters who had never smoked and more than 500 mL for Emergency Medical System (EMS) workers who had never smoked. FEV1 values declined at an average rate of 25 mL per year for firefighters who had never smoked and 40 mL per year for EMS workers who had never smoked. Though values are within the range of 20 to 56 mL per year reported for healthy, nonsmoking men between the ages of 24 and 65, the rate of decline among EMS workers was significantly faster than that for firefighters. This study demonstrated the utility of using longitudinal data from an occupational health surveillance system in disaster epidemiologic research.

Train Derailment in South Carolina

Lesson—sometimes community engagement in the public health and research activities mitigates selection biases. In 2005 a late-night train derailment adjacent to a textile mill in Graniteville, South Carolina released approximately 60 tons of chlorine gas in the town center [42]. State health and environmental officials partnered with the Centers for Disease Control and Prevention to immediately evaluate the health effects. Rapid epidemiologic assessment was initiated while advancing the public health agenda of targeting needed resources to the affected community [43]. Mandatory reporting of persons treated for chlorine-related symptoms accelerated active case finding. Emergency department logs from hospitals in South Carolina and Georgia revealed that acute pulmonary toxicity precipitated nine immediate fatalities and 71 hospitalizations immediately after the train derailment [44]. Though not all were hospitalized, we know of 840 who required medical care [28]. In order to evaluate the urgent long-term health needs following chlorine exposure, a community based participatory research project (CBPR) was initiated and is on-going [20]. This collaboration initially resulted in two public health interventions designed to assess the persistent health effects of those most likely affected and their need of medical assistance, including mental health assessment [28]. In order to evaluate the urgent long-term health needs after the chlorine exposure, the community was engaged within the public health planning and research process. This public health/community collaboration initially resulted in two public health interventions designed to assess the persistent health needs of those most likely affected and their need of medical assistance [45,46], including mental health assessment [28]. As the other disasters have previously illustrated, there is a constant struggle to mitigate selection bias in epidemiologic studies of disaster populations. This is even more of a concern when there is the likelihood of health disparities across race [47] and lack of trust in public officials who some may think should have prevented the EPHD. Distrust is a very normal response because trust has been broken in EPHDs, and such affected communities are often already socio-economically disadvantaged [48]. Therefore it is imperative to fully engage the study community within the scientific approach, building trust, and establishing local credibility prior to initiating any studies [49]. When a community is a partner in the scientific process with a vested interest in the results, they are more likely to participate and comply with the study protocol. We did so from the beginning in Graniteville with many town-hall meetings, outreach to churches and pastors, door-to-door campaigns, collaboration with the local schools and businesses, community information workshops, health fairs and picnics, *etc.* Community engagement successfully overcame the selection bias issue in the Graniteville public health interventions [20]. This community engagement and collaboration has fostered supplemental community based participatory research (CBPR) which is on-going [20]. CBPR maximizes participation rates and minimizes attrition by engaging the participants in the entire research process [49,50,51,52, 53,54,55,56,57,58,59,60,61,62,63,64,65,66,67,68]. Therefore, CBPR strategies are crucial when studying a community impacted by an environmental disaster [69,70,71,72,73,74,75,76,77]. A longitudinal study using CBPR is currently evaluating pulmonary function among a cohort of Graniteville millworkers with baseline pulmonary function tests recorded prior to the chlorine spill. This millworker study will employ a natural experimental design comparing before and after lung health in an interrupted time series. Personal exposure estimates will be generated from plume dispersion modeling [78]. Key successes in evaluating this EPHD were joint consideration of public health needs and epidemiologic research methods, involving the community in short- and long-term assessments and supplemental CBPR, developing a computer model of the plume, leveraging occupational health surveillance data for public health purposes, and accessing existing medical record information to evaluate disease severity within the affected community.

DISCUSSION

The effects of EPHDs are wide-ranging. Lessons learned from the public health evaluation of EPHDs involving contaminant dispersal at the World Trade Center, Bhopal, Chernobyl, and South Carolina provide insights about study design, exposure and outcome assessments, and have implications for evaluating long-term health effects of both recent (e.g., Fukushima) and future disasters.

EPHDs can include hazards with little documented human toxicity data. Therefore, any novel reports of health effects within such exposed populations advance our greater toxicological understanding. But performing epidemiological studies in EPHD populations is not easy. Groups representative of increasing exposure and a similar unexposed group from within the same population [79,80] should be included, even if only using self-reported exposure location data. Public health researchers must judiciously decide upon participant enrollment plans, operational definitions, inclusion/exclusion criteria, and appropriate study designs while minimizing harm and maximizing benefit. Appropriate participant selection, migration patterns, temporal sequence uncertainty, and exposure estimation are major challenges in EPHD studies. However, appropriate exposure assessment during both the early and late phases of an environmental disaster is vital.

During the early phase, identification and quantification of released material and characterization of dispersion patterns helps define the at-risk population, determine potential health risks, and enable exposure estimation [81]. At the WTC and Bhopal, recall of exposure location alone successfully facilitated exposure estimation. During the late phase exposure assessments can further advise public health interventions, identify populations suitable for observational study, and evaluate exposure mitigation strategies. After Chernobyl, exposure measures for individuals were proxied by soil ^{137}Cs measurements and used successfully in a new cohort study [16].

Epidemiological study of discrete communities with relatively localized hazardous exposures can produce significantly fruitful information about the toxicity of the exposure hazard. However, such studies require logistical care because of mistrust among the affected communities which have been violated could result in poor participation, high attrition, volunteer bias, and failed studies [5,82,83,84,85,86,87]. To directly address mistrust concerns, the WTC and Graniteville efforts used community based participatory research related principals (CBPR). In CBPR research investigators partner with the disaster population to identify research questions, plan, implement, evaluate, and disseminate targeted interventions. Scientists must choose to first focus on the public's health needs and the science second when working within EPHDs.

Epidemiological investigations following EPHDs are uniquely poised to accomplish etiologic research [88]. As noted in these case studies, natural experiment designs (e.g., pre/post or interrupted time-series) can be useful when pre-disaster baseline data are available, as with pulmonary function among firemen and EMS workers at the WTC and millworkers in South Carolina. In the case of smaller scale events, statistical power is maximized by using longitudinal designs. When exposure is uniformly nested in a closed longitudinal cohort, each participant serves as their own continuously equivalent control. This design is robust to potential confounding effects of time-invariant covariates (e.g., sex, race), while adjustment may still be needed for time-variant covariates (e.g., season, interviewer). However, because this design assumes uniform and randomly distributed exposure within the study population, selection of appropriate exposure groups requires the same rigor called for in other observational study designs.

Longitudinal cohort studies can be used to determine incident health consequences from an EPHD. In the case of the Chernobyl disaster a longitudinal cohort study was not an initially feasible method for monitoring latent health effects from ionizing radiation exposure. Soviet secrecy was responsible for strategic censorship and data falsification [14]. Therefore, due to the difficulties in obtaining reliable data, there are relatively few conclusive research studies on the non-cancer health effects of ionizing radiation stemming from the Chernobyl disaster despite the many thousands who have been exposed.

Lastly, "ecological epidemiology" studies can provide important insights into hazard exposure *versus*non-hazard exposure causes of latent health effects after EPHDs [29]. The Chernobyl sentinel studies, for example, provide a cost-effective option for evaluating long-term complications and are a nice template for building the arguments for follow-up human health studies.

CONCLUSIONS

EPHDs studies involving dispersal of environmental hazards fill important toxicology knowledge gaps in the latent health effects of these agents. We found several lessons learned from our collective disaster epidemiology work in such EPHDs. These include that sometimes simple study designs are sufficient, public health interventions can facilitate longitudinal studies, "ecological epidemiology" can be a useful alternative to human studies, careful inclusion of an unaffected control population may be better than comparing distance to exposure only, using longitudinal occupational health cohorts can augment population-based studies, and community engagement in the public health and research activities may mitigate selection biases. Furthermore, protocols for immediate and long-term assessment of environmental exposures should be commissioned rapidly in initial response efforts. While difficult methodological challenges exist in health studies of disasters, they are not insurmountable. Discussions of methodological issues should take into consideration both good science and public health practices. Sound scientific research elucidates exposure and response associations, examines the competence of extant response measures, and informs intervention strategies that can be used to evaluate and potentially mitigate long-term EPHD complications.

ACKNOWLEDGEMENTS

The authors would like to acknowledge the assistance of Courtney Mazzoli and Zaina Querishi in the preparation of the manuscript. This research was partially supported by a grant from the National Institute of Environmental Health Sciences (5 R01 ES015532).

REFERENCES

1. Lederman, S.A.; Jones, R.L.; Caldwell, K.L.; Rauh, V.; Sheets, S.E.; Tang, D.; Viswanathan, S.; Becker, M.; Stein, J.L.; Wang, R.Y.; *et al*. Relation between cord blood mercury levels and early child development in a World Trade Center cohort. *Environ. Health Perspect.* 2008, *116*, 1085–1091.
2. Dijkema, M.B.; Grievink, L.; Stellato, R.K.; Roorda, J.; van der Velden, P.G. Determinants of response in a longitudinal health study following the firework-disaster in Enschede, The Netherlands.*Eur. J. Epidemiol.* 2005, *20*, 839–847.
3. Fenig, M.; Cone, D.C. Advancing disaster epidemiology and response: Developing a national disaster-victim database. *Prehosp. Emerg. Care* 2005, *9*, 457–467.
4. Keim, M. Using a community-based approach for prevention and mitigation of national health emergencies. *Pac. Health Dialog.* 2002, *9*, 93–96.
5. Dhara, R.; Dhara, V.R. Bhopal—A case study of international disaster. *Int. J. Occup. Environ. Health*1995, *1*, 58–69.
6. Dhara, V.R. What ails the Bhopal disaster investigation? *Int. J. Occup. Environ. Health* 2002, *8*, 371–379.
7. Cullinan, P.; Acquilla, S.; Dhara, V.R. Respiratory morbidity 10 years after the Union Carbide gas leak at Bhopal: A cross sectional survey. The International Medical Commission on Bhopal. *BMJ*1997, *314*, 338–342.
8. Dhara, V.R.; Acquilla, S.D.; Cullinan, P. Personal exposure and long-term health effects in survivors of the Union Carblde disaster at Bhopal. *Environ. Health Perspect.* 2002, *110*, 487–500.

9. Hudson, J.I.; Pope, H.G.; Glynn, R.J. The cross-sectional cohort study. *Epidemiology* 2005, *16*, 355–359.
10. Yablokov, A.V.; Nesterenko, V.B.; Nesterenko, A.V.; Sherman-Nevinger, J.D. Chernobyl: Consequences of the catastrophe for people and the environment. *Radiat. Prot. Dosim.* 2010, *141*, 97–101.
11. Cardis, E.; Howe, G.; Ron, E.; Bebeshko, V.; Bogdanova, T.; Bouville, A.; Carr, Z.; Chumak, V.; Davis, S.; Demidchik, Y.; *et al.* Cancer consequences of the Chernobyl accident: 20 years on. *J. Radiol. Prot.* 2006, *26*, 127–140.
12. Cardis, E. Current status and epidemiological research needs for achieving a better understanding of the consequences of the Chernobyl accident. *Health Phys.* 2007, *93*, 542–546.
13. Sokolik, G.A.; Ovsiannikova, S.V.; Ivanova, T.G.; Leinova, S.L. Soil-plant transfer of plutonium and americium in contaminated regions of Belarus after the Chernobyl catastrophe. *Environ. Int.* 2004, *30*, 939–947.
14. Yablokov, A.V. Chernobyl's public health consequences: Some methodological problems. *Ann. New York Acad. Sci.* 2009, *1181*, 32–41.
15. Svendsen, E.; Karmaus, W.; Vdovenko, V.; Naboka, M.; Kolpakov, I.; Stepanova, E. Dietary supplementation with clean food improves health following community exposure to ^{137}Cs. *Epidemiology* 2009.
16. Stepanova, E.; Karmaus, W.; Naboka, M.; Vdovenko, V.; Mousseau, T.; Shestopalov, V.M.; Vena, J.; Svendsen, E.; Underhill, D.; Pastides, H. Exposure from the Chernobyl accident had adverse effects on erythrocytes, leukocytes, and, platelets in children in the Narodichesky region, Ukraine: A 6-year follow-up study. *Environ. Health* 2008.
17. Kolpakov, I.Y.; Vdovenko, V.N.; Stepanova, Y.I.; Bazyka, D.A.; Karmaus, W.J.; Svendsen, E.R. Functional state of the respiratory and immune system in children-residents of the radioactive contaminated territories. *Likars'ka sprava/Ministerstvo okhorony zdorov'ia Ukrainy* 2011, *1*, 21–29.
18. Svendsen, E.R.; Kolpakov, I.E.; Stepanova, Y.I.; Vdovenko, V.Y.; Naboka, M.V.; Mousseau, T.A.; Mohr, L.C.; Hoel, D.G.; Karmaus,

W.J. ^{137}Cesium exposure and spirometry measures in Ukrainian children affected by the Chernobyl nuclear incident. *Environ. Health Perspect.* 2010, *118*, 720–725.

19. Sharifkov, V.F.; Koulikova, N.V.; Voropai, L.V. Finding of the Chernobyl Sasakava Health and Medical Cooperation Project: ^{137}Cesium Concenterations among Children around Chernobyl. In Procceding of Fifth Chernobyl Sasakava Medical Cooperation Symposium, Kiev, Ukraine, 14-15 October 1996.
20. Svendsen, E.R.; Kolpakov, I.E.; Stepanova, Y.I.; Vdovenko, V.Y.; Naboka, M.V.; Mousseau, T.A.; Mohr, L.C.; Hoel, D.G.; Karmaus, W.J.J. ^{137}Cesium exposure and spirometry measures in Ukranian children affected by the Chernobyl nuclear incident. *Environ. Health Perspect.* 2010, *118*, 720–725.
21. Gashchak, S.P.; Makliuk, I.A.; Maksimenko, A.M.; Maksimenko, V.M.; Martynenko, V.I.; Chizhevskii, I.V.; Bondar'kov, M.D.; Mousseau, T.A. The features of radioactive contamination of small birds in Chernobyl zone in 2003-2005. *Radiats. Biol. Radioecol.* 2008, *48*, 28–47.
22. Møller, A.P.; Nishiumi, I.; Suzuki, H.; Ueda, K.; Mousseau, T.A. Differences in effects of radiation on abundance of animals in Fukushima and Chernobyl. *Ecol. Indic.* 2013, *24*, 75–81.
23. Møller, A.P.; Barnier, F.; Mousseau, T.A. Ecosystem effects 25 years after Chernobyl: Pollinators, fruit set, and recruitment. *Oecologia* 2012.
24. Moller, A.P.; Mousseau, T.A. Conservation consequences of Chernobyl and other nuclear accidents.*Biol. Conserv.* 2011, *144*, 2787–2798.
25. Mousseau, T.A.; Moller, A.P. Landscape portrait: A look at the impacts of radioactive contaminants on Chernobyl's wildlife. *Bull. Atom. Sci.* 2011, *67*, 38–46.
26. Moller, A.P.; Mousseau, T.A. Efficiency of bio-indicators for low-level radiation under field conditions. *Ecol. Indic.* 2011, *11*, 424–430.
27. Moller, A.P.; Hagiwara, A.; Matsui, S.; Kasahara, S.; Kawatsu, K.; Nishiumi, I.; Suzuki, H.; Ueda, K.; Mousseau, T.A. Abundance of birds in Fukushima as judged from Chernobyl. *Environ. Pollut.*2012, *164*, 36–39.

28. Ginsberg, J.P.; Holbrook, J.R.; Chanda, D.; Bao, H.; Svendsen, E.R. Posttraumatic stress and tendency to panic in the aftermath of the chlorine gas disaster in Graniteville, South Carolina. *Soc. Psychiatr. Psychiatr. Epidemiol.* 2011, *47*, 1441–1448.
29. Møller, A.; Bonisoli-Alquati, A.; Rudolfsen, G.; Mousseau, T. Chernobyl birds have smaller brains.*PLoS One* 2011.
30. Moller, A.P.; Mousseau, T.A.; de Lope, F.; Saino, N. Elevated frequency of abnormalities in barn swallows from Chernobyl. *Biol. Lett.* 2007, *3*, 414–417.]
31. Moller, A.P.; Mousseau, T.A.; Lynn, C.; Ostermiller, S.; Rudolfsen, G. Impaired swimming behaviour and morphology of sperm from barn swallows *Hirundo rustica* in Chernobyl. *Mutat. Res.* 2008, *650*, 210–216.
32. Forum, C. Environmental and Health Consequences of the Chernobyl Accident. In Proceedings of the CHERNOBYL: Looking Back to Go Forward, Vienna, Austria, 6–7 September 2005.
33. Thorpe, L.E.; Friedman, S. Health consequences of the World Trade Center disaster: A 10th anniversary perspective. *J. Am. Med. Assoc.* 2011, *306*, 1133–1134.
34. Balmes, J.R. The World Trade Center collapse: A continuing tragedy for lung health? *Am. J. Respir. Crit. Care Med.* 2006, *174*, 235–236.
35. Lin, S.; Jones, R.; Reibman, J.; Bowers, J.; Fitzgerald, E.F.; Hwang, S.A. Reported respiratory symptoms and adverse home conditions after 9/11 among residents living near the World Trade Center.*J. Asthma* 2007, *44*, 325–332.
36. Moline, J.; Herbert, R.; Nguyen, N. Health consequences of the September 11 World Trade Center attacks: A review. *Cancer Invest.* 2006, *24*, 294–301.
37. Samet, J.M.; Geyh, A.S.; Utell, M.J. The legacy of World Trade Center dust. *N. Engl. J. Med.* 2007,*356*, 2233–2236.
38. Lin, S.; Jones, R.; Reibman, J.; Morse, D.; Hwang, S.A. Lower respiratory symptoms among residents living near the World Trade Center, two and four years after 9/11. *Int. J. Occup. Environ. Health* 2010,*16*, 44–52.
39. Lin, S.; Gomez, M.I.; Gensburg, L.; Liu, W.; Hwang, S.A. Respiratory and cardiovascular hospitalizations after the World

Trade Center disaster. *Arch. Environ. Occup. Health* 2010, *65*, 12–20.

40. Lin, S.; Reibman, J.; Bowers, J.A.; Hwang, S.A.; Hoerning, A.; Gomez, M.I.; Fitzgerald, E.F. Upper respiratory symptoms and other health effects among residents living near the World Trade Center site after September 11, 2001. *Am. J. Epidemiol.* 2005, *162*, 499–507.
41. Aldrich, T.K.; Gustave, J.; Hall, C.B.; Cohen, H.W.; Webber, M.P.; Zeig-Owens, R.; Cosenza, K.; Christodoulou, V.; Glass, L.; Al-Othman, F.; *et al.* Lung function in rescue workers at the World Trade Center after 7 years. *N. Engl. J. Med.* 2010, *362*, 1263–1272.
42. Henry, C.; Belflower, A.; Drociuk, D.; Gibson, J.J.; Harris, R.; Horton, D.K.; Rossiter, S.; Orr, M.; Wright, S.; Wenck, M.A. Public health consequences from hazardous substances acutely released during rail transit—South Carolina, 2005; Selected states, 1999–2004. *J. Am. Med. Assoc.* 2005, *54*, 64–67.
43. Wenck, M.A.; van Sickle, D.; Drociuk, D.; Belflower, A.; Youngblood, C.; Whisnant, M.D.; Taylor, R.; Rudnick, V.; Gibson, J.J. Rapid assessment of exposure to chlorine released from a train derailment and resulting health impact. *Public Health Rep.* 2007, *122*, 784–792.
44. Van Sickle, D.; Wenck, M.A.; Belflower, A.; Drociuk, D.; Ferdinands, J.; Holguin, F.; Svendsen, E.; Bretous, L.; Jankelevich, S.; Gibson, J.J.; *et al.* Acute health effects after exposure to chlorine gas released after a train derailment. *Am. J. Emerg. Med.* 2009, *27*, 1–7.
45. Runkle, J.; Zhang, H.; Karmaus, W.; Brock-Martin, A.; Svendsen, E. An interrupted time series analysis: Prediction of unmet primary health care need for medically vulnerable post-disaster. *Am. J. Publ. Health* 2012. in press.
46. Runkle, J.; Zhang, H.; Karmaus, W.; Brock-Martin, A.; Svendsen, E. Long-term impact of environmental public health disaster on health system performance: Experiences from the Graniteville, South Carolina Chlorine spill. *South. Med. J.* 2012. in press.
47. Davis, J.R.; Wilson, S.; Brock-Martin, A.; Glover, S.; Svendsen, E.R. The impact of disasters on populations with health and health care disparities. *Disaster Med. Public Health Prep.* 2010, *4*, 30–38.

48. Elliott, M.R.; Wang, Y.; Lowe, R.A.; Kleindorfer, P.R. Environmental justice: Frequency and severity of US chemical industry accidents and the socioeconomic status of surrounding communities. *J. Epidemiol. Commun. Health* 2004, *58*, 24–30.
49. Israel, B.A.; Schulz, A.J.; Parker, E.A.; Becker, A.B. Community-based participatory research: Policy recommendations for promoting a partnership approach in health research. *Educ. Health (Abingdon, England)* 2001, *14*, 182–197.
50. Svendsen, E.R.; Whittle, N.C.; Sanders, L.; McKeown, R.E.; Sprayberry, K.; Heim, M.; Caldwell, R.; Gibson, J.J.; Vena, J.E. GRACE: Public health recovery methods following an environmental disaster.*Arch. Environ. Occup. Health* 2010, *65*, 77–85.
51. Avery, R.C.; Wing, S.; Marshall, S.W.; Schiffman, S.S. Odor from industrial hog farming operations and mucosal immune function in neighbors. *Arch. Environ. Health*. 2004, *59*, 101–108.
52. Higgins, D.L.; Metzler, M. Implementing community-based participatory research centers in diverse urban settings. *J. Urban Health* 2001, *78*, 488–494.
53. Horowitz, C.R.; Arniella, A.; James, S.; Bickell, N.A. Using community-based participatory research to reduce health disparities in East and Central Harlem. *Mt. Sinai J. Med.* 2004, *71*, 368–374.
54. Israel, B.A.; Parker, E.A.; Rowe, Z.; Salvatore, A.; Minkler, M.; Lopez, J.; Butz, A.; Mosley, A.; Coates, L.; Lambert, G.; *et al.* Community-based participatory research: Lessons learned from the Centers for Children's Environmental Health and Disease Prevention Research. *Environ. Health Perspect.* 2005, *113*, 1463–1471.
55. Israel, B.A.; Schulz, A.J.; Parker, E.A.; Becker, A.B. Review of community-based research: Assessing partnership approaches to improve public health. *Ann. Rev. Public Health* 1998, *19*, 173–202.
56. Lantz, P.M.; Viruell-Fuentes, E.; Israel, B.A.; Softley, D.; Guzman, R. Can communities and academia work together on public health research? Evaluation results from a community-based participatory research parternship in Detroit. *J. UrbanHealth* 2001, *78*, 495–507.

57. Leung, M.W.; Yen, I.H.; Minkler, M. Community based participatory research: A promising approach for increasing epidemiology's relevance in the 21st century. *Int. J. Epidemiol.* 2004, *33*, 499–506.

58. Lipscomb, H.J.; Argue, R.; McDonald, M.A.; Dement, J.M.; Epling, C.A.; James, T.; Wing, S.; Loomis, D. Exploration of work and health disparities among black women employed in poultry processing in the rural south. *Environ. Health Perspect.* 2005, *113*, 1833–1840.

59. Minkler, M. Community-based research partnerships: Challenges and opportunities. *J. Urban Health*2005, *82*, ii3–ii12.

60. Minkler, M.; Blackwell, A.G.; Thompson, M.; Tamir, H. Community-based participatory research: Implications for public health funding. *Am. J. Public Health* 2003, *93*, 1210–1213.

61. Minkler, M.; Vasquez, V.B.; Tajik, M.; Petersen, D. Promoting environmental justice through community-based participatory research: The role of community and partnership capacity. *Health Educ. Behav.* 2006, *109*.

62. O'Fallon, L.R.; Dearry, A. Commitment of the National Institute of Environmental Health Sciences to community-based participatory research for rural health. *Environ. Health Perspect.* 2001, *109*, 469–473.

63. O'Fallon, L.R.; Dearry, A. Community-based participatory research as a tool to advance environmental health sciences. *Environ. Health Perspect.* 2002, *110*, 155–159.

64. Parker, E.A.; Israel, B.A.; Williams, M.; Brakefield-Caldwell, W.; Lewis, T.C.; Robins, T.; Ramirez, E.; Rowe, Z.; Keeler, G. Community action against asthma: Examining the partnership process of a community-based participatory research project. *J. Gen. Intern. Med.* 2003, *18*, 558–567.

65. Schulz, A.J.; Krieger, J.; Galea, S. Addressing social determinants of health: Community-based participatory approaches to research and practice. *Health Educ. Behav.* 2002, *29*, 287–295.

66. Schulz, A.J.; Parker, E.A.; Israel, B.A.; Becker, A.B.; Maciak, B.J.; Hollis, R. Conducting a participatory community-based survey for a community health intervention on Detroit's east side. *J. Public Health Manag. Pract.* 1998, *4*, 10–24.

67. Vasquez, V.B.; Minkler, M.; Shepard, P. Promoting environmental health policy through community based participatory research: A case study from Harlem, New York. *J. Urban Health* 2006, *83*, 101–110.
68. Wing, S.; Wolf, S. Intensive livestock operations, health, and quality of life among eastern North Carolina resident. *Environ. Health Perspect.* 2000, *108*, 233–238.
69. Viswanathan, M.; Ammerman, A.; Eng, E.; Garlehner, G.; Lohr, K.N.; Griffith, D.; Rhodes, S.; Samuel-Hodge, C.; Maty, S.; Lux, L.; *et al.* Community-based participatory research: Assessing the evidence. *Evid. Rep. Technol. Assess. (Summ.)* 2004, *18*, 1–8.
70. Heath, S.E.; Kenyon, S.J.; Zepeda Sein, C.A. Emergency management of disasters involving livestock in developing countries. *Rev. Sci. Tech.* 1999, *18*, 256–271.
71. Krajeski, R.L.; Peterson, K.J. "But she is a woman and this is a man's job": Lessons for participatory research and participatory recovery. *Int. J. Mass Emerg. Disasters* 1999, *17*, 123–130.
72. Bolin, R.; Stanford, L. The Northridge earthquake: Community-based approaches to unmet recovery needs. *Disasters* 1998, *22*, 21–38.
73. Bailey, C.; Convery, I.; Mort, M.; Baxter, J. Different public health geographies of the 2001 foot and mouth disease epidemic: "Citizen" *versus* "professional" epidemiology. *Health Place* 2006, *12*, 157–166.
74. Yehuda, R.; Bryant, R.; Marmar, C.; Zohar, J. Pathological responses to terrorism.*Neuropsychopharmacology* 2005, *30*, 1793–1805.
75. Moore, S.; Daniel, M.; Linnan, L.; Campbell, M.; Benedict, S.; Meier, A. After Hurricane Floyd passed: Investigating the social determinants of disaster preparedness and recovery. *Fam. Commun. Health* 2004, *27*, 204–217.
76. Group, T.W. "Gold standard" for remediation of wtc contamination. *New Solut.* 2004, *14*, 199–217.
77. Farquhar, S.; Dobson, N. Community and university participation in disaster-relief recovery: An example from Eastern North Carolina. *J. Commun. Pract.* 2004, *12*, 203–217.

78. Chemtob, C.M.; Nakashima, J.P.; Hamada, R.S. Psychosocial intervention for postdisaster trauma symptoms in elementary school children: A controlled community field study. *Arch. Pediatr. Adolesc. Med.* 2002, *156*, 211–216.

79. Reed, D.; Feigley, C.; Svendsen, E. Exposure estimation technology for large environmental chemical releases: Feasibility of an atmospheric plume dispersion model without microenvironmental weather data. *Epidemiology* 2011, *22*, S105–S106.

80. Cullinan, P. Epidemiological assessment of health effects from chemical incidents. *Occup. Environ. Med.* 2002, *59*, 568–572.

81. Bongers, S.; Janssen, N.A.; Reiss, B.; Grievink, L.; Lebret, E.; Kromhout, H. Challenges of exposure assessment for health studies in the aftermath of chemical incidents and disasters. *J. Expo. Sci. Environ. Epidemiol.* 2008, *18*, 341–359.

82. Bertazzi, P.A. Industrial disasters and epidemiology. A review of recent experiences. *Scand. J. WorkEnviron. Health* 1989, *15*, 85–100.

83. Dhara, V.R. What ails the Bhopal disaster investigations? (And is there a cure?). *Int. J. Occup. Environ. Health* 2002, *8*, 371–379.

84. Dhara, V.R.; Dhara, R. The union carbide disaster in Bhopal: A review of health effects. *Arch. Environ. Health* 2002, *57*, 391–404.

85. Grievink, L.; van der Velden, P.G.; Yzermans, C.J.; Roorda, J.; Stellato, R.K. The importance of estimating selection bias on prevalence estimates shortly after a disaster. *Ann. Epidemiol.* 2006, *16*, 782–788.

86. Huizink, A.C.; Smidt, N.; Twisk, J.W.R.; Slottje, P.; Smid, T. Epidemiological disaster research: The necessity to include representativesamplesoftheinvolveddisasterworkers. Experience from the epidemiological study air disaster Amsterdam-ESADA. *J. Epidemiol. Commun. Health* 2006, *60*, 887–889.

87. Van den Berg, B.; van der Velden, P.; Stellato, R.; Grievink, L. Selective attrition and bias in a longitudinal health survey among survivors of a disaster. *BMC Med. Res. Methodol.* 2007.

88. Dominici, F.; Levy, J.I.; Louis, T.A. Methodological challenges and contributions in disaster epidemiology. *Epidemiol. Rev.* 2005, *27*, 9–12.

Chapter 9

Method of Searching for Earthquake Disaster Evacuation Routes Using Multi-Objective GA and GIS

Yuichiro Shimura and Kayoko Yamamoto

Graduate School of Information Systems, University of Electro-Communications, Tokyo, Japan

ABSTRACT

This study treats the determination of routes for evacuation on foot in earthquake disasters as a multi-objective optimization problem, and aims to propose a method for quantitatively searching for evacuation routes using a multi-objective genetic algorithm (multi-objective GA) and GIS. The conclusions can be summarized in the following three points. 1) A GA was used to design and create an evacuation route search algorithm which solves the problem of the optimization of earthquake disaster evacuation routes by treating it as an optimization problem with multiple objectives, such as evacuation distance and evacuation time. 2) In this method, goodness of fit is set by using a Pareto ranking method to determine the ranking of individuals based on their relative superiorities and inferiorities. 3) In this method, searching for evacuation routes based on the information on present conditions allows evacuation routes to be derived based on present building and road locations. Further, this method is based on publicly available information; therefore, obtaining geographic information similar to that of this study enables this method to be effective regardless of what region it is applied to, or whether the data regards the past or the future. Therefore, this method has high degree of spatial and temporal reproducibility.

INTRODUCTION

Based on experiences in the Great Hanshin Earthquake (1995) and the Great East Japan Earthquake (2011), in recent years in Japan, reduction of the effects of natural disasters through self-help and mutual help is being emphasized, and effort is being put into the formulation of evacuation plans by independent disaster prevention organizations and the like. In 2011, in "Towards Reconstruction" [1], the Reconstruction Design Council in Response to the Great East Japan Earthquake offered the opinion that self-help which takes the form of the idea of "escape" from disaster should be made the basis of the approach to reducing the effects of natural disasters, and in order to enable this, measures concerning more intangible aspects should be emphasized. These measures include thorough disaster prevention education conducted using mutual aid and the provision of hazard maps by public bodies. In order to "escape" quickly and safely from disaster, it is necessary to formulate a clear-cut evacuation plan-specifically, to create evacuation routes. Examples of existing efforts to formulate evacuation plans through self-help and mutual help are bousai machi-aruki activities (disaster prevention activities that involve walking around a neighborhood to check which places are safe to evacuate to in the event of a disaster, and which are dangerous), disaster simulation exercises and so on.

A representative example of the latter is DIG (Disaster Imagination Game), developed by Komura et al. (1997) [2]. However, a common feature in the creation of evacuation routes through these efforts is that since the main objectives are inspiring disaster prevention awareness and educating people about disaster prevention, the routes are not evaluated quantitatively. Therefore, the evacuation plans formulated are influenced by the subjective points of view of the creators, and uncertainty remains as to whether they would be of practical use in a disaster.

Based on the above-mentioned background, this study treats the determination of routes for evacuation on foot in earthquake disasters as a multi-objective optimization problem, and aims to propose a method for quantitatively searching for evacuation routes using a multi-objective genetic algorithm (multi-objective GA). Further, the method of this study is applied to the region studied, and multiple evacuation route candidates that are derived are visualized on a digital map using GIS and evaluated. This study aims to use the method to derive routes in order to efficiently provide information for natural disaster reduction efforts carried out through self- help and mutual help—that is, to efficiently provide information for the formulation of evacuation plans by individuals and independent disaster prevention organizations.

RELATED WORK

This study is related to 1) Evacuation routes; 2) Optimal solution searches using GA; 3) Spatial analysis using GIS; and 4) Multi-objective optimization problems. The study by Liu et al. (2006) [3] is an example of research related to (1). Liu et al. used flood damage simulation by a spatiotemporal geographic information system (DiMSIS) and the Dijkstra method, which is a method for finding the shortest route, and developed an algorithm for deriving evacuation routes that can respond to changes in the situation over time in a flood disaster. Further, Takeuchi et al. (2002) [4] imagined evacuation from a tsunami in an earthquake disaster and analyzed evacuation routes which took into account road blockage caused by the collapse of buildings and overall evacuation distance. Ichikawa et al. (2001) [5] took into account the fact that blockage of roads by collapsed buildings makes evacuation difficult, and estimated the danger on routes to evacuation areas for each road. They then used a Monte Carlo simulation to generate the road blockage situation, and considered the location of evacuation areas based on the road blockage situation on the shortest routes.

As examples of research related to (2), Inoue et al. (2013) [6] derived the optimal location of tourism-related facilities in urban tourist areas, and Asou et al. (2007) [7] derived the optimal location of evacuation sites in disasters. These studies demonstrated the superiority of the multipoint search feature of GA in the combinational optimization problem of facility location evaluation. As an example of research concerning (3), Koarai et al. (2007) [8] superposed a map showing the physical characteristics of an area of land with information about building damage caused by past earthquakes on a GIS, and analyzed the superposed map and information in order to examine the relationship between earthquake-induced ground disaster and the physical characteristics of land. They demonstrated the effectiveness of spatial analysis which employs GIS.

In the area of research regarding multi-objective optimization problems, there are many studies which have conducted optimal solution searches using a multi-objective GA. These searches apply the GA multipoint search feature. Hara et al. (2006) [9] viewed route searching by car navigation systems as a multi-objective optimization problem. They proposed a method in which a multi-objective GA was used to optimize independent objective functions for route length, estimated journey time, and degree of driving comfort. They demonstrated the effectiveness of the method by comparing it with existing methods. Kinoshita et al. (2008) [10] used a multi- objective GA to generate sightseeing routes which took into account the preferences of each member in a group, and demonstrated the superiority of searching by multi-objective GA in the case of optimizing multiple conflicting objective functions.

Based on the results of the above-mentioned related research, this study demonstrates uniqueness and value in the field of disaster research in that it aims to use a multi-objective GA to quantitatively search for evacuation routes, in order to provide effective information for evacuation plan formulation carried out using self-help and mutual help methods, which are more direct forms of natural disaster reduction help; that is, it aims to use a multi-objective GA to quantitatively search for evacuation routes in order to provide effective information for evacuation plan formulation by individual evacuees and regional disaster prevention organizations.

FRAMEWORK AND METHOD

In Section 4, the challenges of avoiding each of the various causes of obstruction of evacuation are set as the objectives in the evacuation route optimization, and in order to solve the multi-objective optimization problem of optimizing these objectives, an evacuation route search algorithm which employs a GA is designed and created. In Section 5, road network data for the district studied is created using GIS, and information about obstructions which occur during evacuation is added to this data for each road. In Section 6, an evacuation route search algorithm is applied to the road network that was created, evacuation starting points and destinations are set, and an evaluation experiment is performed. Further, the convergence of evaluation values for each objective is examined, and whether or not an effective search is being conducted or not is investigated. In Section 7, the parameters which resulted in the most effective search in Section 6 are applied to the GA, and an evacuation route search is conducted again. Further, the routes derived are visualized in GIS, and superposed with information for each road, such as information on blockage probability. This allows a visual evaluation to be performed, and superior routes are proposed. Additionally, Esri Inc.'s ArcGIS Ver. 10.1 is used as the GIS. The GIS is used to create road network data, perform spatial analysis, and visualize evacuation route search results.

The district chosen for study was the northern part of Sumida City (Figure 1), which is the Tokyo Metropolitan district that would be the most vulnerable in the event of a disaster, according to the "District-based assessment of vulnerability to earthquake disaster (No. 7)" [11] , published by the Tokyo Metropolitan Government. This district has a high concentration of wooden dwellings, and many roads with a narrow width; therefore, it is thought that there are many factors which would hinder evacuation in the event of a disaster. Thus, the effectiveness of the method of this study can be demonstrated well in this district.

EVACUATION ROUTE SEARCH METHOD EMPLOYING MULTI-OBJECTIVE GA

Design of Evacuation Route Search Algorithm

Python 2.7 was used as the programming language for the design of the evacuation route search algorithm. Figure 2 shows the flow of the evacuation route search algorithm, and an outline of each operation is given below.

Figure 1: Location of district studied.

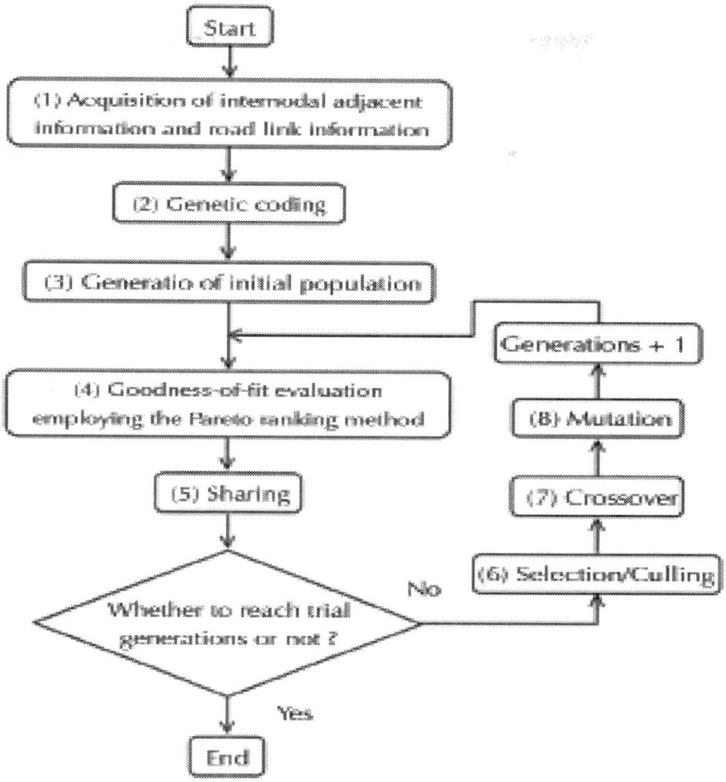

Figure 2: Flow of evacuation route search algorithm.

- Acquisition of internodal adjacent information and road link information

Change the CSV data of the node-related adjacent information and the CSV data of the road link information into list form, and input the data into the GA.

- Genetic coding

Determine how each of the routes, which are individuals, will be expressed in the GA.

- Generation of initial population

Randomly generate a number of routes equal to the population size (the number of individuals which are to be in the population) and form the initial population.

- Goodness-of-fit evaluation employing the Pareto ranking method

Using the evaluation values for each objective (road information), calculate goodness-of-fit of each route based on the Pareto ranking method.

- Sharing

To the goodness-of-fit calculated in (4), add revisions which take into account similarities between individuals, and create a new goodness-of-fit.

- Selection/Culling

Based on the goodness-of-fit to which revisions were added in (5), determine which routes to preserve for the next generation. As a selection method, use a combination of roulette-wheel selection and the preservation of Pareto strategy (elite preservation strategy in a single-objective optimization GA).

- Crossover

Select two routes from the population to be parents, and by crossing these over at a crossover point, generate two routes which are children.

- Mutation

Select one route from the population, and once again randomly generate routes at and after the mutation point, in order to generate new routes.

Creation of Evacuation Route Search Algorithm

Acquisition of Internodal Adjacent Information and Road Link Information

- Creation of adjacency list

Transform CSV data which contains the internodal adjacency relationships into adjacency list form, and input it into the GA. Concerning the adjacency list, it is given a triple structure, node numbers are made to correspond with the adjacency list index, and adjacency information is stored. As adjacency information, node numbers adjacent to each node and link numbers which link nodes are stored.

- Creation of road link information list

A link list, formed by transforming CSV data containing road link information into a list, is input into the GA. When this is done, link numbers are made to correspond with the link list index, and road information is stored.

Genetic Coding

This study deals with routes; therefore, the order of node numbers, which forms routes, is an issue. Therefore, in this study, genetic coding is performed using path expression. Path expression is a method in which genes are node numbers, and routes are expressed by lists arranged in the order in which node numbers are passed through. Path expression can be said to be the most natural method. Further, since lists express routes as is, path expression is characterized by the fact that decoding from genotype to phenotype is not necessary. Gene length varies with the number of nodes passed through; therefore, it is variable.

Generation of Initial Population

A number of the individuals coded in the previous section that is equal to the population size number N is randomly generated to form the initial population. The population is represented as a list with a dual structure. In order to prevent routes from looping, nodes which have already been passed through are not used. Further, individuals are generated based on the adjacency list created in Section 4.2.1, and this ensures that only feasible solutions—that is, only routes that are connected—are generated. In this method, the genetic operations of selection/culling, crossing, and mutation are performed on the population generated in the manner described above.

Evaluation of Goodness-of-Fit, Using the Pareto Ranking Method

- Determination of evaluation values regarding each objective

In this study, the four objectives of reduction of evacuation distance increase in the probability of arriving at the destination which takes road blockages into account, avoidance of hazard caused by fire, and reduction of evacuation time were taken into account in evacuation route optimization. Therefore, firstly, for all individuals, evaluation values for each objective (distance, arrival probability, degree of fire hazard in roads, and estimated evacuation time) were defined using the undermentioned Equations (1) to (3) and (6).

- Evaluation value for distance

The evaluation value for distance was found as the total evacuation distance to the destination, using the following Equation (1).

$$f1_i = \sum_{j=0}^{k-1} l_{ij} \tag{1}$$

$f1_i$: Total distance for individual X_i [m];

l_{ij}: Distance for the j-th link which individual X_i passes through [m];

k: Total number of links which individual X_i passes through;

- Evaluation value for arrival probability

Concerning the evaluation value for arrival probability: For each link, the probability of being able to pass through the link was calculated based on the probability of road blockage for the link. The probability that it would be possible to pass through all road links in the route was the arrival probability. The evaluation value for arrival probability was found as an evaluation value for these probabilities.

$$f2_i = \prod_{j=0}^{k-1} \left(1 - P_{ij}\right) \tag{2}$$

$f2_i$: Arrival probability for individual X_i;

P_{ij}: Probability of blockage for j-th link that individual X_i passes through;

1- P_{ij}: Probability that j-th link that individual X_i passes through will be passable;

k: Total number of links that individual X_i passes through.

- Evaluation value for degree of fire hazard in roads

The evaluation value for the degree of fire hazard in roads is the sum total of the degree of fire hazard in roads for each link, and is found using Equation (3).

$$f3_i = \sum_{j=0}^{k-1} d_{ij} \tag{3}$$

$f3_i$: Total degree of fire hazard in roads for individual X_i;

d_{ij}: Degree of fire hazard in roads for j-th link that individual X_i passes through;

k: Total number of links that individual X_i passes through.

- Evaluation value for estimated evacuation time

The evaluation value for estimated evacuation time is the sum total of the periods of time an evacuee spends passing through links, and is calculated based on the distance, road width, and estimated evacuation population in the road for each link. Based on material from the Urban Disaster Prevention Office of the City Bureau of the Ministry of Construction (1997) [12] and the Director-General for Disaster Management in the Cabinet Office (2005) [13] , an equation for deriving the road width that allows a certain number of evacuees to pass through in an hour is given below.

$$\text{(Road width)} = \text{(Number of evacuees)}/\text{(Evacuation population density)}/\text{(Walking speed)}/\text{(Total evacuation time)} \quad (4)$$

Here, the evacuation population density indicates the number of people that can evacuate per 1 m^2. Generally, one person per square meter is used as the evacuation population density during a disaster. That is equivalent to the density on a crowded train station platform. Further, usually 2 km/h is used as the walking speed during a disaster. This is based on the opinion that it is advisable for designated large-area evacuation sites to be located such that the time it takes to evacuate to them is within an hour and the evacuation distance is within 2 km. With reference to the above-mentioned equation for deriving road width, in the present method, the transit time t_j for link j is defined by Equation (5), given below.

$$t_j = \frac{l_j}{s} + \frac{\sum_{j=0}^{j} u_j}{\rho \cdot s \cdot w_j} \quad (5)$$

t_j: Time required to pass through link j [h];

l_j: Distance for link j [m];

u_j: Estimated evacuation population in road for link j [P];

$\sum_{j=0}^{j} u_j$: Sum total of estimated evacuation population in roads for links passed through before reaching link j [P];

s: Walking speed [m/h] (=2000 m/h);

ρ: Evacuation population density [P/m²] (=1 P/m²);

w_j: Road width for link j [m].

Here, the second term on the right side of Equation (5) expresses the evacuation time, found based on the equation for deriving road width. Further, concerning the estimated evacuation population, when a link is passed through, the estimated evacuation population at that link increases cumulatively. Based on the above, the total evacuation time for individual X_i is defined by the Equation (6), given below, and is an evaluation value.

$$f4_i = \sum_{j=0}^{k-1} t_j = \sum_{j=0}^{k-1} \left(\frac{l_j}{s} + \frac{\sum_{j=0}^{j} u_j}{\rho \cdot s \cdot w_j} \right) \tag{6}$$

$f4_i$: Total evacuation time for individual X_i [h];

k: Total number of links that individual X_i passes through.

The Pareto ranking method is used based on the evaluation values for each of these objectives, to determine objective functions for calculating the goodness of fit of individuals.

- Goodness of fit evaluation employing the Pareto ranking method

By using the Pareto ranking method based on evaluation values for each objective defined in (1), a rank which expresses the superiority or inferiority of each individual is determined. In the case below, the individual X_i is defined as being "dominated by" X_j.

$$f1_i \geq f1_j \wedge f2_i \leq f2_j \wedge f3_i \geq f3_j \wedge f4_i \geq f4_j \tag{7}$$

f1: Total evacuation distance;

f2: Probability of arriving at destination;

f3: Sum total of degree of fire hazard in roads;

f4: Total evacuation time.

That is, in the case where individual X_i has a longer distance, a lower probability of arriving, a higher degree of fire hazard in roads, and a longer total evacuation time than individual X_j, individual X_i is defined as being "dominated by" individual X_j. Based on this definition, the rank of individual X_i (expressed by "$R(X_i)$") when individual X_i is dominated by a number of individuals equal to n_i is determined by Equation (8), given below.

$$R(X_i) = 1 + n_i \tag{8}$$

An individual for which R = 1 holds true indicates a Pareto optimal solution, which is a solution that is not inferior to any other solution. This rank is used to set objective functions, as in Equation (9) below. Equation (9) is set such that the higher the rank (the closer to 1), the better the goodness of fit.

$$F_i = \frac{\sum_{j=0}^{N-1} R(X_j)}{R(X_i)} \tag{9}$$

Thus, objective functions are set based on ranks found using the Pareto ranking method.

Sharing

In the method of this study, sharing in the representational space is conducted. If sharing in the objective space was applied to the method, and routes with similar evaluation values for each objective existed within the population, those routes would be regarded as being similar. However, because routes are focused on in this study, even if evaluation values are similar, it is possible that in the representational space the nodes which form the routes may be completely different; therefore, the routes cannot be regarded as being similar. Due to this reason, in the method of this study, similarity as routes is taken into account, and unique sharing in the representational space is applied. The specific method of sharing in this study is outlined below.

- The number of nodes which overlap between individuals X_i and X_j are counted. However, the evacuation starting point and destination are not counted.
- The total number of nodes which form X_i and X_j is counted. However, the evacuation starting point and destination are not counted.

- The proportion of nodes that overlap between X_i and X_j (expressed as "$r(X_i,X_j)$") is found. The proportion of overlapping nodes is equivalent to the distance d between individuals in sharing in the objective space.
- Steps (1) to (3) are conducted between all individuals in the population, and the proportion of nodes which overlap between individuals is found for every individual in relation to every other individual.
- For each individual, a niche number, which is the number of individuals which are similar to that individual, is calculated. The niche number m_{X_i} of individual X_i is defined by Equation (10), given below.

$$m_{X_i} = \sum_{j=0}^{N-1} s\left(r\left(X_i, X_j\right)\right) \tag{10}$$

s(r): Sharing function;

$r(X_i,X_j)$: Proportion of nodes that overlap between individuals X_i and X_j;

N: Population size.

The sharing function s(r) is defined by Equation (11), given below.

$$s(r) = \begin{Bmatrix} 1 & (\text{if } r \geq r_{\text{share}}) \\ 0 & (\text{otherwise}) \end{Bmatrix} \tag{11}$$

r_{share}: Sharing rate,$0 \leq r_{share} \leq$

In this study, the sharing rate r_{share} is equivalent to the sharing radius σ of sharing in the objective space. Arbitrary values of from 0 to 1 are given to r_{share}. That is, in the case where the proportion of nodes which overlap between individuals is equal to or greater than the sharing rate, those individuals are regarded as being similar, and their niche numbers are counted.

- By dividing the goodness of fit F of each individual by the niche number found for each individual, a new goodness of fit share F which takes into account similarity between individuals is created. The goodness of fit of the individual X_i, indicated by share F_i, is expressed by Equation (12), given.

$$\text{share}F_i = \frac{F_i}{m_{X_i}} \tag{12}$$

By conducting selection and culling based on the value for the goodness of fit share F_i for which the sharing described above has been conducted, the diversity of individuals in the population is maintained, and a more global search, which is important in seeking a set of Pareto optimal solutions, can be conducted. Concerning the effectiveness of the sharing conducted in the method of this study, in Section 6.1 an evaluation experiment is conducted, and the dispersion of individuals within the population is checked to verify whether or not sharing has been conducted effectively.

Selection and Culling

In the method of this study, the strategy of preserving Pareto optimal solutions and roulette-wheel selection are used in combinationas selection and culling operations. Each operation is described below. Selection other than that for individuals whose preservation for the next generation has been determined using the strategy of preserving Pareto optimal solutions is performed using roulette-wheel selection. The probability of selection for individuals in roulette-wheel selection is indicated by pselect. It is based on the goodness of fit for which sharing has been conducted, which is indicated by shareF, and is determined by Equation (13), given below.

$$\text{pselect}_i = \frac{\text{share}F_i}{\sum_{j=0}^{N-1} \text{share}F_j} \tag{13}$$

Based on the probability of selection for each individual, which is determined by Equation (13), roulette- wheel selection is repeated until the number of next-generation individuals reaches the population size N. By determining probability selection based on goodness of fit which, through the use of sharing, takes into account similarity between individuals, as in Equation (13), a next generation population in which individuals are dispersed evenly and there is no prevalence of similar individuals is generated.

Crossover

The crossover method of this study is described below. As the crossover method, one-point crossover is used.

- Two individuals are randomly selected from the population to be parent individuals. In the case that the two individuals are formed using exactly the same nodes, the individual generated will be the same as the parent individuals; therefore, crossover is not performed with such individuals.
- From nodes which overlap between the two individuals (not including the evacuation starting points and destination points), one point is randomly selected as a crossover point. If there are no overlapping nodes, crossover is not conducted.
- Nodes at and after the crossover point of the parent individuals are swapped, and two child individuals are generated.
- If a route loops in a child individual, the loop is removed.

The above-described operations are performed based on the crossover probability C, and generated child individuals are added to the population.

Mutation

The mutation method used in the method of this study is described below.

- One individual is randomly selected from the population.
- One point is randomly selected from nodes which form the selected individual, and used as a mutation point. However, evacuation starting point and destination points are not selected as mutation points.
- From the mutation point node, a route which links to the destination point is randomly generated. Route generation is performed based on adjacency lists, similarly to initial population generation, and only feasible solutions, or unbroken routes, are generated.
- The part of the individual selected in (1) that is at and beyond the mutation point is replaced by the randomly generated route, resulting in the generation of a new route.

- If a route generated in (4) loops, the loop is removed.

The above-described operations are performed based on the mutation rate M, and generated individuals are added to the population.

CREATION OF ROAD INFORMATION FOR DISTRICT STUDIED

Evacuation Rules

Evacuation Procedure

The Tokyo Metropolitan Government [14] has two-stage evacuation as a fundamental rule for evacuation in times of disaster, and Sumida City [15] [16] also follows this system. In two-stage evacuation, as the first stage of evacuation after the outbreak of a disaster, people evacuate to a temporary gathering place, and ascertain the damage. As temporary gathering places, elementary schools, junior high schools, parks and so on are designated by each neighborhood association/ residents' association. In the case where the damage is ascertained and the temporary gathering place is judged to be dangerous, as the second stage of evacuation, evacuees evacuate to a designated large-area evacuation site. The Tokyo Metropolitan Government designates places which are roughly 10 ha or more in area as large-area evacuation sites [14]. It is thought that places with an area of 10 ha or more are not affected by radiant heat from fires. Meanwhile, in cases where damage from fire spreads on a large scale after the outbreak of a disaster, and it is judged dangerous to evacuate to a temporary evacuation place, evacuees evacuate directly to a large-area evacuation site. This is referred to as direct evacuation. This study focuses on two-stage evacuation and direct evacuation, and attempts to optimize evacuation routes for each of these methods.

Allocation of Designated Large-area Evacuation Sites

Each district is assigned a set large-area evacuation site. In the district studied, the 3-chome Kyoujima area has the highest total degree of danger, according to Sumida City [16]. The total degree of danger (the sum of the building collapse risk and the degree of fire hazard) takes into account the degree of difficulty in carrying out activities during a disaster. The Tokyo Metropolitan Government [14] has designated that the large-area evacuation site for the 3-chome Kyoujima area is the Tokyo Metropolitan Bunka 1-chome residential zone (Toei Bunka 1-chome jūtaku ittai). Therefore, the 3-chome Kyoujima area was selected from northern Sumida City as a focus of this study, and this study conducts searches for evacuation routes whose destination is the Tokyo Metropolitan Bunka 1-chome residential zone.

Creation of Road Information Using GIS

Collection and Processing of Data Used

Table 1 show the data used in this study and how the data was processed. Information on the probability of total building collapse, sorted by ground area and building type, from "District-based assessment of vulnerability to earthquake disaster (No. 7)" [11] published by the Tokyo Metropolitan Government was used in calculating the road blockage probability and adding it to the road network. However, data on the year each building was constructed is fixed asset tax roll data which falls into the category of personal information, and it is difficult for individuals to obtain. Therefore, in this study, the case of the worst damage possible was imagined, and it was assumed that all wooden buildings were constructed prior to 1970. Further, based on "District-based assessment of vulnerability to earthquake disaster (No. 7)" [11], a uniform total building collapse rate of 50.99% was used. Further, in this study, crowding on roads due to population concentration is taken into account as a chief cause of obstacles to evacuation. The night and day populations sorted by administrative address units such as block, subdivision and aza (the smallest administrative unit in Japan)were considered (Table 1), and it was found that in the 3-chome Kyoujima area the nighttime population was greater than the daytime population; therefore, the nighttime population was used to calculate the estimated population in roads.

Creation of Road Network Data

In this study, the road blockage probability was calculated based on the positional relationship between the debris width buffer caused by roadside wooden buildings, and the road centerline located in front of that. Therefore, the road centerline was extracted from land use data, and was used to create node data which showed intersections and crossroads, and link data which showed roads which connected the intersections and crossroads. Road network data was created by further integrating these sets of data. Figure 3 shows the road network created as described above, and road information was added to this data.

Table 1: Data used and data processing method

Data used	Data processing method
Land use survey for FY2012 (Bureau of Urban Development, Tokyo Metropolitan Government)	Road network generation
Building use survey for FY2012 (Bureau of Urban Development, Tokyo Metropolitan Government) Total building collapse rate, sorted by ground area and building type (Bureau of Urban Development, Tokyo Metropolitan Government: "District-based assessment of vulnerability to earthquake disaster (No. 7)")	Road blockage probability calculation and its addition to road network
Land use survey for FY2012 (Bureau of Urban Development, Tokyo Metropolitan Government)	Road width measurement

Vulnerability to fire, sorted by block/subdivision (Bureau of Urban Development, Tokyo Metropolitan Government: "District-based assessment of vulnerability to earthquake disaster (No. 7)")	Addition of degree of fire hazard to road network
Data on nighttime population, taken from data on the night and day populations, sorted by block, subdivision, aza, and other administrative units (Statistics Bureau, Ministry of Internal Affairs and Communications: "Heisei 22-nen Kokusei Chousa [2010 National Census]") Building use survey for FY2012 (Bureau of Urban Development, Tokyo Metropolitan Government)	Calculation of estimated population in roads and its addition to road network

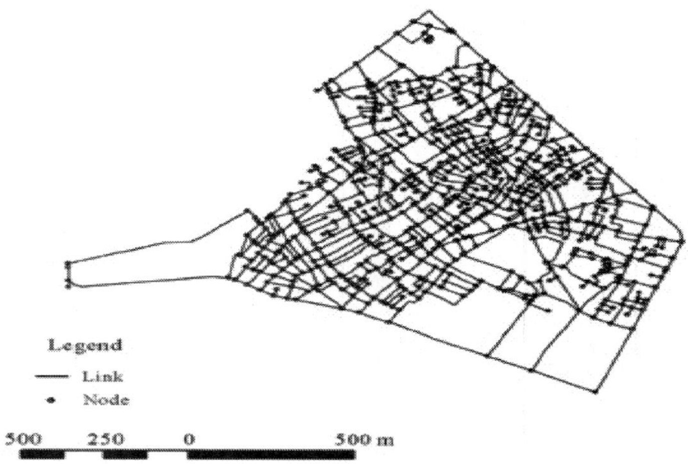

Figure 3: Road network.

Road Blockage Probability Calculation and Its Addition to the Road Network

Road blockage probability was added to the road network data in Figure 3 as an attribute. This study takes into account two types of blockage: 1) Single blockage, defined as being when a road is blocked by the collapse of a wooden building on only one side of a road, and 2) Combined blockage, defined as being when a road is blocked by the collapse of wooden buildings on both sides of a road. Here, in the case of (1) single blockage, when the debris width produced by a collapsed building intersects with a building shape located on the other side of the road, the road located between these buildings is regarded as being blocked. Further, in the case of (2) combined blockage, when the debris widths produced by collapsed buildings on both sides of the road intersect, the road located between these collapsed buildings is regarded as being blocked. Concerning road blockage probability for the case of (1) single blockage, since the road will be blocked by the collapse of one wooden building on one side of the road, the blockage probability is regarded as being the same as the probability that a wooden building will collapse. Meanwhile, in the case of (2) combined blockage, since the road is blocked when buildings on both the left and right sides of the road collapse, the blockage probability is found by multiplying the collapse probabilities for buildings on the left and right sides of the road. Below, methods for calculating road blockage probability and adding it to the road network are described.

- Roadside building extraction and classification

Roadside buildings, which are a primary factor in road blockage, were extracted. First, a buffer from the road centerline was created on the right side only, and the spatial search function of the GIS was used to extract buildings which intersected with the buffer. Similarly, next a buffer from the road centerline was created on the left side only, and buildings which intersected with the buffer were extracted. By following these steps, buildings were classified into those located on the right side of the road and those located on the left side of the road.

- Extraction of roadside wooden buildings and addition of probability of collapse

Using the GIS attribute search function, wooden buildings were extracted from both the buildings on the right side and the buildings on the left side. Further, probability of collapse was added as an attribute to each of the wooden buildings.

- Extraction of pairs of buildings which would cause blockage

1. **Case of single blockage**

A buffer with a radius equal to the width of debris which would be produced from a wooden building on the right side of the road if it collapsed was created, and the GIS spatial search function was used to extract any building on the left side of the road which intersected with the buffer. The same was done for the left side of the road, and pairs of buildings which would cause single blockage were extracted. Concerning the outflow of debris that occurs when a building collapses, the case of maximum damage was imagined, and it was assumed that debris would flow out evenly in all directions. Further, concerning the debris width, with reference to Ichikawa et al. (2001) [5] , the debris width was found using Equation (14), given below. Concerning the number of floors in buildings, reference was made to data on the present building use situation. As the average floor height, a uniform average floor height of three meters was used.

$$(\text{Debris width}) = (\text{Number of floors}) \times (\text{Average floor height}) \tag{14}$$

2. **Case of combined blockage**

Debris width buffers were created for wooden buildings on both the left and right sides of the road. The GIS spatial search function was used to extract left-side debris buffers which intersected with right-side debris buffers and right-side debris buffers which intersected with left-side debris buffers. These steps were performed to extract buildings on the right side and buildings on the left side which would cause combined blockage.

- Generation of road section lines

With reference to Yoshikawa et al. (2003) [17], section lines (lines which connected the center points of buildings which would cause road blockage) were created. Using the GIS XY To Line tool, line data was generated for lines which connected the center points of buildings which would cause road blockage (which were extracted in (3) above), and this data was used as section lines. XY coordinates which were the center points of buildings were found using the GIS geometry arithmetic function. Section lines were generated to suit three types of case—the case of single blockage caused by the collapse of wooden buildings on the right side of the road, the case of single blockage caused by the collapse of wooden buildings on the left side of the road, and the case of combined blockage. Further, blockage probability which suited the type of blockage was added for each case.

- Addition of blockage probability to road network

Using the GIS spatial coupling tool, to each road link which intersected with the three types of section line generated in (4), blockage probability (indicated by p_k) for each of the points on section lines (indicated by k) was added.

- Calculation of road blockage probability

Through the above operations, to each road link, blockage probabilities (indicated by p_k) for points on section lines (these points were indicated by k) were added. The probability P_s that all points k on the section lines of each road link would be passable was calculated using the blockage probability p_k, and is expressed by Equation (15) (Takeuchi et al. (2002) [4]).

$$P_s = \prod (1 - p_k) \tag{15}$$

Therefore, the road link blockage probability P_b was found using the Equation (16) below:

$$P_b = 1 - P_s = 1 - \prod (1 - p_k) \tag{16}$$

Using the GIS field arithmetic function, the above-described calculations were performed on a road link data attribute table, and blockage probability was calculated for each road link. Figure 4 shows the road network to which road probability blockage for each link, calculated using the above operations, was added. From Figure 4, it can be seen that blockage probability is high in the northeastern part.

Measurements of Road Width

Road width was measured based on road edge line data and road centerlines (road links) created in Section 5.2.1. Road width is necessary when calculating fire hazard and evacuation time. The method for measuring road width is described below.

- Based on road edge line data, buffers with a radius of 0.5 m were generated, and road links which intersected with the buffers were extracted using the GIS spatial search function. This enabled roads with a width of less than 1 m to be extracted.

- A width of 1 m (the case where the buffer radius is 0.5 m) was given to the extracted road links as an attribute.
- Road links extracted in (2) were excluded from the overall body of road link data.
- Once again, buffers were generated from road edge line data. When performing this step, the buffer radius was a value obtained by adding 0.5 m to the buffer radius set in (1). Further, using the GIS spatial search function, the group of road links which did not include roads with a width of less than 1 m (which was created in (3)) was superposed with the buffers, and road links which intersected with the buffers were extracted. This step enabled roads with a width greater than or equal to 1 m and less than 2 m to be extracted.

Steps (1) to (4) were repeated, increasing the buffer radius by 0.5 m each time. Through this, road link data was extracted for widths every meter up to a maximum width. Finally, the road link data extracted for widths every meter was integrated using the GIS merge function to create one set of road link data containing the addition of width information for every road link. Figure 5 shows a road network to which road width measured as described above has been added.

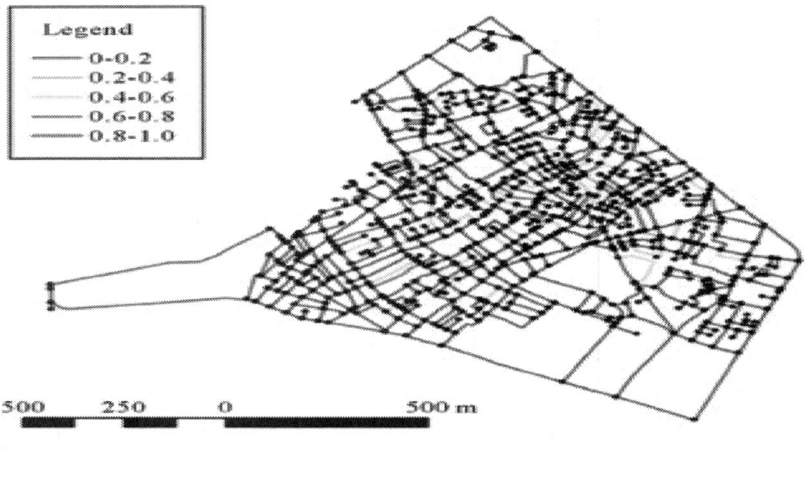

Figure 4: Road network to which road probability blockage has been added.

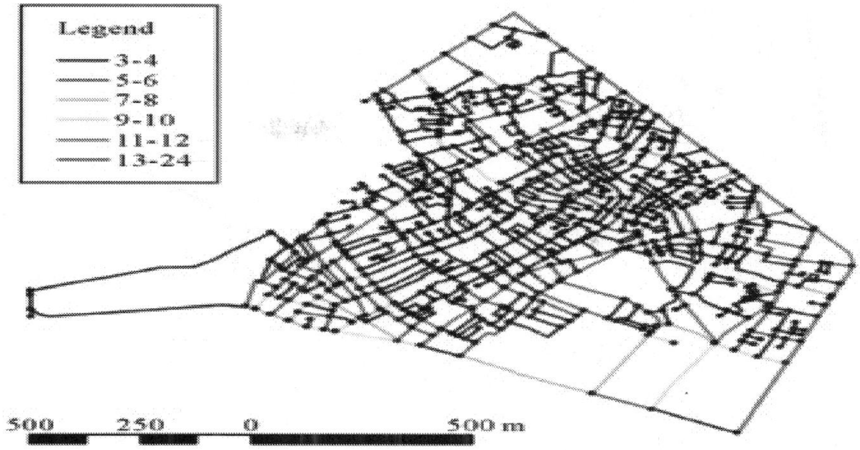

Figure 5: Road network to which road width has been added.

Addition of Degree of Fire Hazard to Road Network

The degree of fire hazard for each block and chome* (* a numbered subdivision of an area in an address) from "District-based assessment of vulnerability to earthquake disaster (No. 7)" [11] was added as the hazard due to fire for each road link. The method for doing so is described below.

- The degree of fire hazard for each block and chome was added as an attribute to block and chome polygon data taken from data on the present land use situation.
- Using a GIS spatial coupling tool, degree of fire hazard was added to road links which intersected with each block and chome polygon. The Great Hanshin Earthquake clearly showed that roads with a width equal to or greater than twelve meters have the effect of interrupting the spread of fire. Therefore, roads with a width of twelve meters or more were assigned a fire hazard degree of zero. Concerning road width data, the values measured in Section 5.2.3 were used.
- The degrees of fire hazard added to road links in step (2) were constant within each block/chome. Therefore, the degree of fire hazard did not differ between the road links in a block/chome.

This meant that no matter what road in the block/chome was passed through, the degree of fire hazard would be the same. However, for evacuation in the case where a fire has broken out, speedy evacuation to roads with a wide width which will not suffer damage from fire or the spread of fire is important; therefore, evacuation distance is thought to be closely related. Here, based on the idea that "The longer the evacuation distance travelled in a block/chome with a high degree of fire hazard, the more likely it is that fire hazard will be encountered", as a new indicator showing the degree of hazard due to fire in a road, a value found by multiplying the degree of fire hazard with the road link distance was used as the degree of fire hazard in a road. Further, the degree of fire hazard in roads for each road was calculated using the GIS field arithmetic function in a road link attribute table. Doing this enabled the degree of fire hazard in roads within a block/chome to be varied.

Figure 6 shows the road network to which the degree of fire hazard in roads created in the above manner was added.

Calculation of Estimated Evacuation Population within Road and Its Addition to the Road Network

The estimated population of evacuees which would occur in each road at the time of a disaster outbreak was calculated, and was added to road links as an attribute. Estimated population is used in the GA when calculating evacuation time for a route. Below, methods for calculating the estimated population within a road and adding it to the road network are described

- Using the GIS spatial search function, based on data on present use of buildings which intersected with block/chome polygon data taken from data on the present land use situation, for each block/chome, the buildings in the block/chome were extracted.
- Using the GIS attribute search function, from the data on buildings extracted in step (1), only the single or detached houses and apartment buildings or housing complexes with many nighttime evacuees were extracted.

- Using Equation (17), estimated populations which matched the total floor space of each residence were assigned. Calculation for Equation (17) was performed using the GIS field arithmetic function in attribute fields for building data. Figure 7 shows the estimated evacuation population for each residence, calculated as described above.

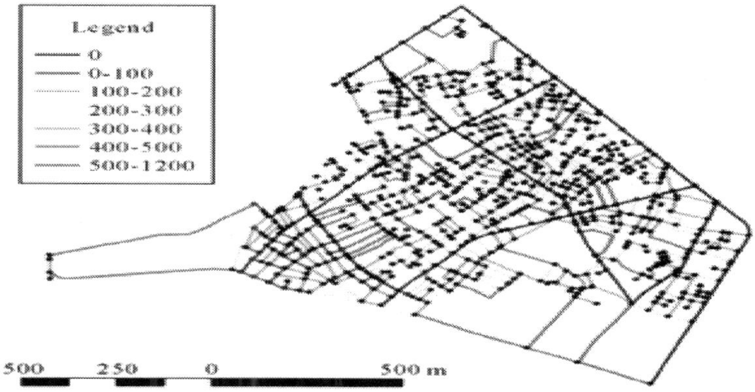

Figure 6: Road network to which degree of fire hazard in roads has been added.

Figure 7: Estimated evacuation population for each residence.

$$(\text{Estimated evacuation population in Building } i) = (\text{Nighttime population of relevant block}/\textit{chome}) \times \frac{(\text{Total floor space of Building } i)}{(\text{Total combined floor space of residential sections in relevant block}/\textit{chome})} \quad (17)$$

- Using the GIS spatial coupling tool, the estimated evacuation population of each residence was assigned to the road link closest to the residence.

Figure 8 shows a road network to which estimated evacuation population in the road calculated as described above was added.

Output of Road Link Data

All road information for the road network created in Sections 5.2.1 to 5.2.6 was integrated into a single GIS attribute table and output. The method of doing so is described below.

- Using the GIS dissolve function, all road information was summarized based on road link IDs. This enabled the distance (Figure 3), road blockage probability (Figure 4), road width (Figure 5), degree of hazard due to fire in road (Figure 6), and estimated evacuation population in road (Figure 8) for each road link to be summarized into a single GIS attribute table.
- The GIS tool for exporting feature attributes to ASCII was used to output the road information in step (1) to a CSV file.

Road link data output as described above is input to the GA as road information when route search is performed.

EVALUATION OF EVACUATION ROUTE SEARCH ALGORITHM

In this section, an evaluation experiment conducted in order to demonstrate the validity of the method of this study is described. In Section 6.1, in order to demonstrate the validity of the unique sharing method of the method of this study, whether or not effective sharing was conducted is verified. This is done by checking the distribution of solutions. At the same time, by considering the change in solutions derived that occurs when the sharing rate r_{share} is changed, an appropriate r_{share} value is clarified. In Section 6.2, by considering the change in solutions in the case where the number of objectives is changed, and the convergence of solutions, whether or not multi-objective optimization was conducted appropriately is examined. Further, at the same time, GIS is used to visualize derived routes, and a visual evaluation is conducted.

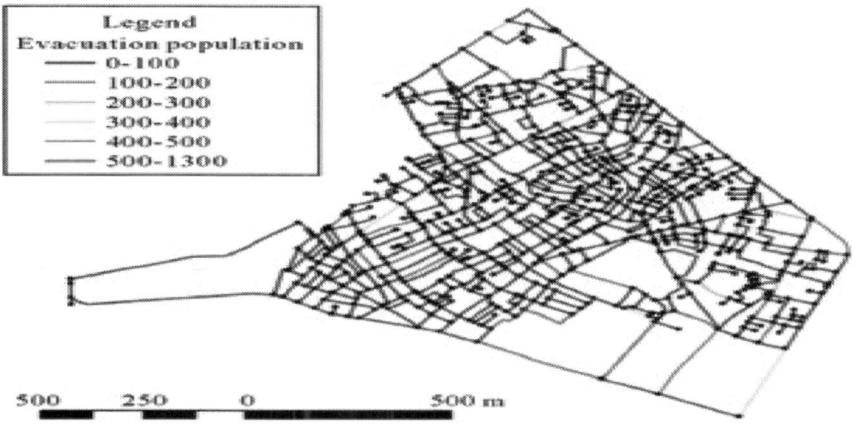

Figure 8: Road network to which estimated evacuation population in roads has been added.

In the evaluation experiment in this section, in order to compare derived routes, the same evacuation starting point and the same evacuation destination point were used throughout the experiment. Specifically, as shown in, Figure 9 in the evaluation experiment, the evacuation starting point was Point A, and the destination was Point B, which is a node adjacent to the large-area evacuation site.

Sharing Evaluation Experiment

In this section, in order to examine the effects of the unique sharing of this study, the distribution of individuals within the population for the case where sharing is not conducted and the case where sharing is conducted and the sharing rate r_{share} is changed are compared. Doing this enables verification of whether or not the objective of sharing, which is "the maintenance of diversity in solutions within the population", was achieved successfully. In this section, optimization is conducted using two objectives for which distribution of individuals is easy to grasp visually (evacuation distance and arrival probability).

GA Parameters

Table 2 shows GA parameters for the evaluation experiment concerning sharing. In order to enable comparison of the effects of sharing in each case from (1) to (6), derivation of solutions was carried out with all parameters except for r_{share} being the same for every case.

Sharing Evaluation Experiment Results and Discussion

Table 3 shows evaluation values for the Pareto optimal solutions derived from individuals in the population of the final generation for each of Cases (1) to (6) in Table 2.

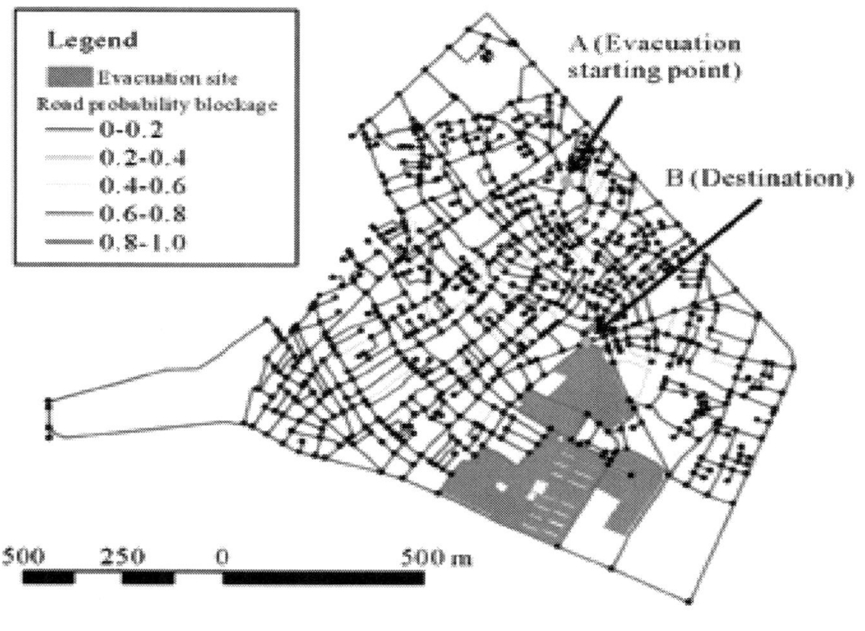

Figure 9: Evacuation starting point and destination in evaluation experiment.

Table 2: GA parameters in sharing evaluation experiment

Case	(1)	(2)	(3)	(4)	(5)	(6)
Sharing rate: r_{share}	0.0	0.2	0.4	0.6	0.8	1.0
Population size: N	100	100	100	100	100	100
Number of trial generations: G	1000	1000	1000	1000	1000	1000
Crossover rate: C [%]	100	100	100	100	100	100
Mutation rate: M [%]	1	1	1	1	1	1

Table 3: Evaluation values for Pareto optimal solutions derived from distribution of individuals in the population of the final generation

Route	(1)		(2)		(3)		(4)		(5)		(6)	
	Evacuation distance [m]	Arrival prob ability	Evacuation distance[m]	Arrival probability	Evacuation distance [m]	Arrival prob ability	Evacuation distance [m]	Arrival probability	Evacuation distance [m]	Arrival probability	Evacuation distance [m]	Arrival probability
1	543.5	0.0051	543.5	0.0051	543.5	0.00 51	543.5	0.0051	543.5	0.0051	543.5	0.0051
2	559.9	0.0110	559.9	0.0110	559.9	0.01 10	559.9	0.0110	559.9	0.0110	559.9	0.0110
3	583.8	0.0190	583.8	0.0190	583.8	0.01 90	583.8	0.0190	583.8	0.0190	583.8	0.0190
4	598.3	0.1000	598.3	0.1000	598.3	0.10 00	598.3	0.1000	598.3	0.1000	598.3	0.1000
5	614.7	0.2400	614.7	0.2400	614.7	0.24 00	614.7	0.2400	614.7	0.2400	614.7	0.2400
6	-	-	678.0	0.7400	654.7	0.36 00	654.7	0.3600	654.7	0.3600	654.7	0.3600
7	-	-	-	-	678.0	0.74 00	678.0	0.7400	678.0	0.7400	666.7	0.4900
8	-	-	-	-	-	-	-	-	-	-	678.0	0.7400

It can be seen that in the case where sharing was not conducted, the population is dominated by individuals which are similar. Therefore, a global search was not conducted, and the search for Pareto optimal solutions was also not conducted sufficiently. The smallest number of Pareto optimal solutions was derived in this case-five. In Case (2), where r_{share} = 0.2, because sharing was conducted, compared to Case (1) there are more types of individual in the population, and further, six Pareto optimal solutions were derived. This is thought to be due to the fact that because individuals in the population were diverse, the number of types of parent individual in the genetic operations of crossover and mutation increased. That is, due to the number of types of parent individual increasing, child individuals generated based on those parent individuals also became diverse, and therefore a more global search was possible. This is also thought to be the reason why more Pareto optimal solutions were derived in Cases (3) to (6), with seven being derived in each of Cases (3) to (5), and eight being derived in Case (6). Further, comparing Cases (3) to (6), it can be seen that individuals in the population are most diverse in Case (6), for which r_{share} = 1.0.

r_{share} = 1.0 indicates the case where sharing which lowers the goodness of fit of individuals formed with exactly the same nodes is performed in the population. In the method of this study, when the two individuals that are parents are formed with exactly the same nodes, crossover operations are not performed. This is because even if the parents were crossed, child individuals which were exactly the same as the parents would be generated. By lowering the goodness of fit of individuals with exactly the same structure through sharing for which r_{share} = 1.0, more diverse individuals are selected in the population through selection and culling, and therefore crossover can be performed effectively. Through this, since crossover allows a greater variety of child individuals to be generated, a greater variety of individuals exists within the population. Due to the above-mentioned reasons, it is thought that in the method of this study, sharing for which r_{share} = 1.0 was the most effective sharing.

Due to these reasons, it was confirmed that the sharing in this study had the effect of maintaining the diversity of individuals within the population, and consequently a global search is possible; therefore, it is likely that a greater number of Pareto optimal solutions will be found. Further, it was confirmed that in the method of this study, sharing is most effective when r_{share} = 1.0. Therefore, r_{share} = 1.0 is used in this method, as it is considered to be the best sharing rate.

Experiment to Evaluate Multi-objective Optimization

This section describes an evaluation experiment which was carried out in order to verify whether or not multi- objective optimization of evacuation routes was carried out appropriately by the method of this study. The GA parameters in the evaluation experiment were as follows. Sharing rate: r_{share}= 1.0; Population size: N = 100; Number of trial generations: G = 1000; Crossover rate: C [%] = 100; and Mutation rate: M [%] = 1. Table 4 summarizes evaluation values for each objective for routes derived, for which evacuation distance, arrival probability, degree of hazard due to fire in the road, and evacuation time were optimized. Twenty routes were derived as Pareto optimal solutions.

Table 4: Evaluation values for routes optimized for evacuation distance, arrival probability, degree of hazard due to fire in the road, evacuation time and total evacuation population

Derived route	Evacuation distance [m]	Arrival probability	Degree of hazard due to fire in the road	Evacuation time [h]	Total evacuation population [p]
1	543.5	0.0052	1839.5	0.43	150
2	559.9	0.0110	1921.3	0.46	158
3	561.8	0.0020	1794.9	0.45	172
4	583.8	0.0190	2040.8	0.49	185
5	598.4	0.1100	2113.6	0.56	222
6	600.0	0.0170	1525.1	0.47	149
7	600.2	0.0430	2122.6	0.53	173
8	608.8	0.0320	1569.4	0.51	165
9	614.7	0.2400	2195.4	0.61	230
10	616.6	0.0430	2069.1	0.60	244
11	619.4	0.0720	2218.8	0.39	81
12	623.2	0.0350	1413.3	0.48	187
13	625.4	0.0130	1424.2	0.47	149

14	629.6	0.0350	1171.3	0.44	203
15	632.0	0.0640	1457.6	0.51	165
16	638.5	0.0640	1215.5	0.47	161
17	654.8	0.3600	1799.3	0.62	221
18	678.0	0.7400	1687.5	0.60	198
19	684.5	0.7400	1445.4	0.56	259
20	1074.9	0.4900	1018.3	0.72	221

Of these, Routes 1, 11, 19 and 20 are shown in Figure 10. Further, Figure 11 and Figure 12 show results for derived routes superposed with road width and estimated evacuation population for each residence, and derived routes superimposed with the estimated evacuation populations in roads and for each residence, respectively.

Concerning the derived routes, including Routes 1, 18, 19 and 20, the routes are almost the same as in the case when evacuation distance, arrival probability, and degree of hazard due to fire in the road were optimized; however, a new route, Route 11, which minimizes evacuation time, was derived. Route 11 does not have the shortest evacuation distance, but it has the shortest evacuation time. Route 1, the shortest route, has a total evacuation population of 150 people, while Route 11 has a total evacuation population of 81 people. Further, Figure 12 clearly shows that while Route 1 passes through roads with a large estimated evacuation population, Route 11 passes through roads with a relatively small estimated evacuation population. Further, Figure 11 clearly shows that while Route 1 mainly passes through roads with a width of three to four meters, Route 11 mainly passes through roads with a width of seven to nine meters. Therefore, it is thought that evacuation time for Route 1 is prolonged due to crowding in the roads that is a result of a large evacuation population, and furthermore, roads with a narrow width. In contrast, Route 11 has a small estimated evacuation population, and wide roads were selected for it; therefore, it is considered that there is hardly any prolongation of evacuation time due to crowding in the roads for Route 11. Thus, it was confirmed that by adding evacuation time as an objective in this manner, new routes which take evacuation time into account can be generated.

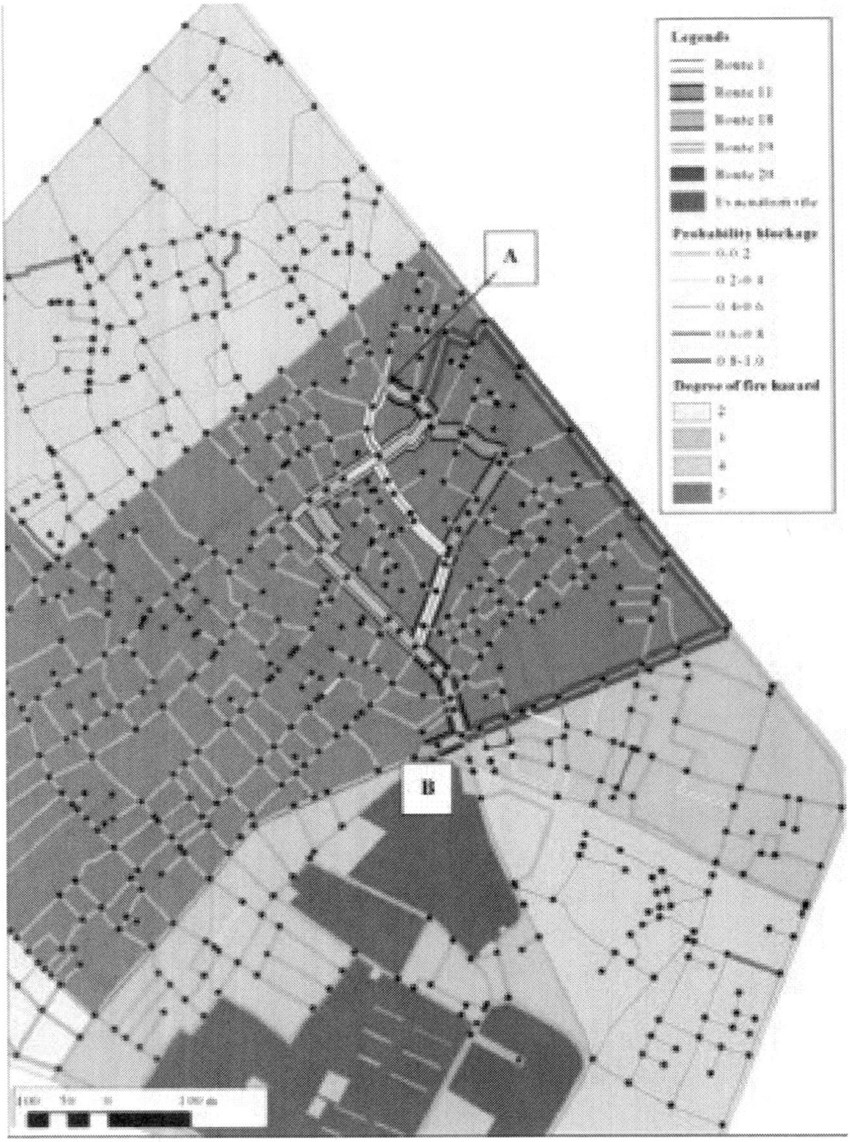

Figure 10: Routes optimized for evacuation distance, arrival probability, degree of hazard due to fire in the road and evacuation time.

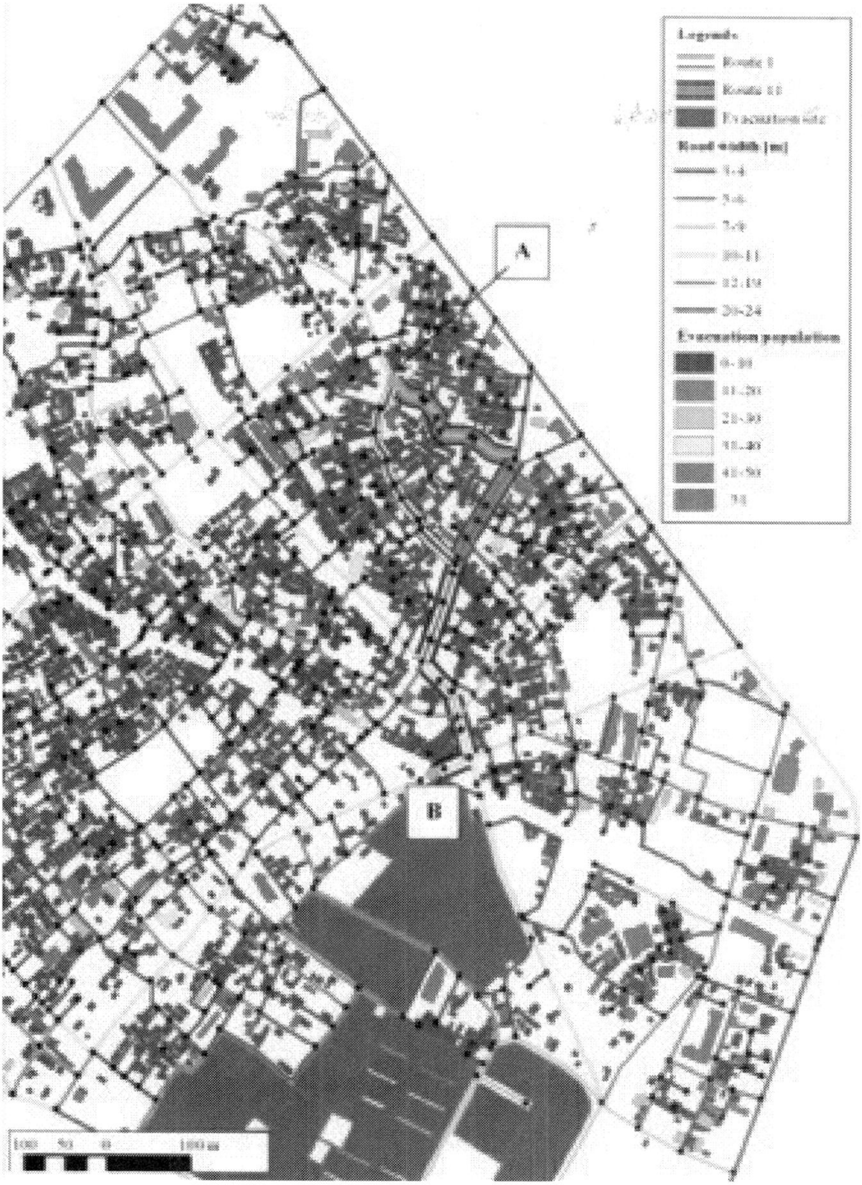

Figure 11: Derived routes, road width and estimated evacuation population for each residence.

Figure 12: Derived routes, estimated evacuation populations in roads and for each residence.

The result that is clear from above is that as the number of objectives was increased, multiple routes whose evaluation values for each objective were intrade-off relationships with each other were obtained. These routes were Pareto optimal solutions, meaning that in order to improve an evaluation value for one of the objectives, an evaluation value for one of the other objectives would have to be degraded. Therefore, it is considered that multi-objective optimization was carried out appropriately. Thus, in the method of this study, multiple evacuation route candidates are derived in a single search as Pareto optimal solutions; therefore, from these candidates, multiple evacuation routes which envisage what will happen in a time of disaster can be selected.

Convergence of Evaluation Values for Each Objective

In order to verify whether or not multi-objective simultaneous optimization was effectively carried out, convergence of evaluation values for each objective in a four-objective optimization was examined. The GA parameters in the evaluation experiment were as follows. Sharing rate: r_{share} = 1.0; Population size: N = 100; Number of trial generations: G = 1000; Crossover rate: C [%] = 100; and Mutation rate: M [%] = 1. Figure 13 shows the convergence of evaluation values for each generation derived. From Figure 13 it can be seen that average values in the populations converged in the 600 meter region for evacuation distance, in the 2000 region for degree of hazard due to fire in the road, and in the 0.5 hour region for evacuation time.

However, concerning arrival probability, it can be seen that while the average value for arrival probability of individuals within the population improved overall compared to the initial generation with the repeated creation of generations, a clear convergence could not be confirmed, compared with the other three objectives. The fact that evacuation distance is intimately connected with all three of the other objectives is considered a reason for this. Evacuation distance itself is an evaluation value that is determined by distance, and degree of hazard due to fire in the road is an evaluation value obtained by multiplying the degree of hazard due to fire per block/chome by the distance of each road. Further, evacuation time is an evaluation value obtained by using evacuation distance and walking speed to find the regular transit time, and then adding extra time to that due to crowding. In the case where there is only a small influence from crowding, evacuation time is strongly dependent on evacuation distance. Due to this, it is thought that the more the evacuation distance for individuals in the population converges on a small value, the more evacuation distance, degree of hazard due to fire in the road, and evacuation time will improve. Meanwhile, since arrival probability is not dependent on evacuation distance, it will not improve if the evacuation distance for individuals in the population becomes shorter. Nevertheless, it can be seen that evaluation values for each objective for individuals in the population improved with the addition of each generation, and it is considered that multi-objective simultaneous optimization was carried out adequately.

PROPOSAL OF EVACUATION ROUTE CANDIDATES IN THE DISTRICT STUDIED

Selection of Representative Evacuation Starting Points and Evacuation Destination Points for Each District

Points in each area for which evacuation was considered to be most difficult based on the total degree of hazard (which takes into account the degree of difficulty of conducting activities in times of disaster) were selected as representative evacuation starting points. Further, evacuation routes are proposed for two types of evacuation method in this study—the case of two-stage evacuation, and the case of direct evacuation.

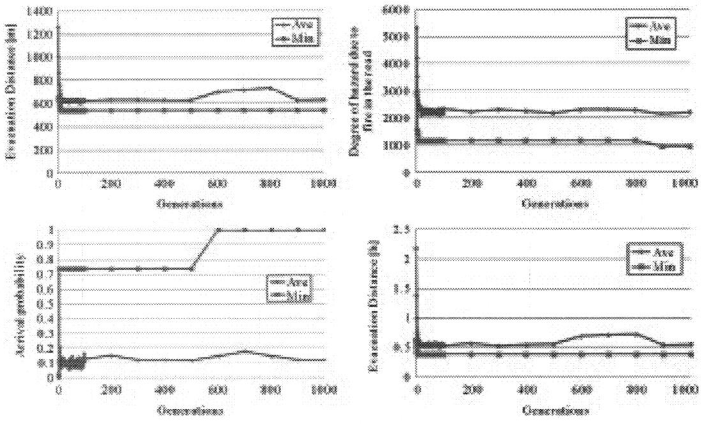

Figure 13: Convergence of each evaluation value with each generation derived.

Therefore, two points were selected as evacuation destination points—a temporary gathering place, which is the first destination point, and a large-area evacuation site, which is the second destination point. The temporary gathering place is a place designated by Sumida City [16] , the location of the evacuation starting point, and the large-area evacuation site is a place designated by the Tokyo Metropolitan Government [14] . Specifically, as shown in Figure 14, Point A was used as the evacuation starting point. Point A is located in 3-chome Kyoujima, an area for which the total degree of hazard (which takes into account the degree of difficulty of conducting activities in times of disaster) is high. Further, the temporary gathering place for the residents of Point A is Karatachi Children's Playground; therefore, Point B, a point adjacent to the playground, was used as the first destination. Point C was used as the second destination. Point C is adjacent to the Tokyo Metropolitan Bunka 1-chome residential zone (Toei Bunka 1-chome jūtaku ittai), which is a large-area evacuation site.

Figure 14: Evacuation starting point and destinations.

GA Parameters

The GA parameters set for route search were as follows. Sharing rate: r_{share} = 1.0; Population size: N = 150; Number of trial generations: G = 1000; Crossover rate: C [%] = 100; and Mutation rate: M [%] = 2.

Examples of Evacuation

Evacuation Routes in the Case of Two-stage Evacuation

- from evacuation starting point to temporary gathering place

As routes from Point A, the evacuation starting point, to Point B, the neighborhood of the temporary gathering place, only the two routes shown in Table 5 and Figure 15 were derived as Pareto optimal solutions.

Route 2 has the shortest distance; however, because it involves passing through roads with a high probability of blockage, it has a low arrival probability of 36%. In contrast to this, Route 1 detours to avoid roads with a high blockage probability, and has an arrival probability of 100%. Further, because Route 1 is not an unnecessarily indirect route, no large differences can be observed between Route 1 and Route 2 for evacuation distance, degree of hazard in the road due to fire, and evacuation time.

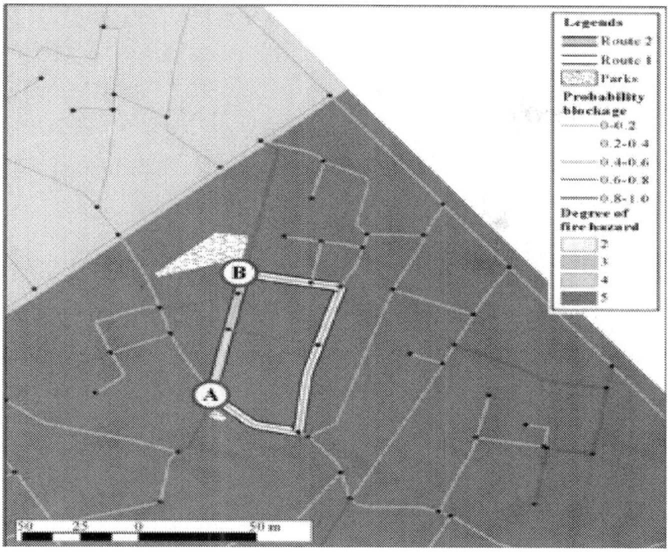

Figure 15: Evacuation routes from evacuation starting point to temporary gathering place.

Table 5: Evaluation values for evacuation routes from evacuation starting point to temporary gathering place

Derived route	Evacuation distance [m]	Arrival probability	Degree of hazard due to fire in the road	Evacuation time [h]
1	164.3	1.00	821.5	0.140
2	69.4	0.36	347.2	0.042

Accordingly, Route 1, which has a high arrival probability, is thought to be suitable as an evacuation route, and therefore was chosen as a preference solution.

- Route from temporary gathering place to large-area evacuation site

As routes from Point B, the neighborhood of the temporary gathering place, to Point C, the neighborhood of the large-area evacuation site, thirty-two routes were derived as Pareto optimal solutions. Table 6 shows evaluation values for the derived routes. Of the derived routes, Routes 27 and 30 are shown in Figure 16. Further, Figure 17 and Figure 18 show results for preference routes superposed with the estimated evacuation populations in roads and for each residence, and derived routes superposed with road width, respectively.

Table 6: Evaluation values for derived routes

Derived route	**Evacuation distance [m]**	**Arrival probability**	**Degree of hazard due to fire in the road**	**Evacuation time [h]**
1	613.0	0.0019	2186.7	0.51
2	629.4	0.0041	2268.5	0.55
3	631.2	0.0007	2142.1	0.54
4	653.3	0.0070	2388.0	0.57
5	665.3	0.0190	2448.3	0.63
6	667.8	0.0390	2460.9	0.65
7	669.4	0.0063	1872.4	0.56
8	669.6	0.0150	2469.8	0.62
9	679.9	0.1100	2521.2	0.71
10	681.7	0.0430	2530.2	0.69
11	683.5	0.0076	2403.8	0.67
12	684.2	0.0870	2542.7	0.71
13	686.0	0.0150	2416.3	0.68
14	688.9	0.0260	2566.0	0.48
15	692.6	0.0130	1760.6	0.56
16	696.2	0.2400	2603.0	0.77
17	698.1	0.0430	2476.6	0.75
18	699.1	0.0130	1518.5	0.52
19	700.9	0.0720	2626.3	0.54
20	711.2	0.0230	1853.6	0.61
21	717.7	0.0230	1611.6	0.56

22	724.2	0.1300	2146.5	0.70
23	736.3	0.3600	2206.8	0.77
24	747.5	0.2700	2034.7	0.68
25	753.9	0.2700	1792.7	0.63
26	759.5	0.7400	2095.0	0.74
27	766.0	0.7400	1853.0	0.69
28	1006.8	0.1800	1627.4	0.65
29	1011.7	0.1800	1651.7	0.64
30	1029.7	1.0000	572.1	0.59
31	1034.5	1.0000	596.4	0.58
32	1106.7	0.0870	389.8	0.61

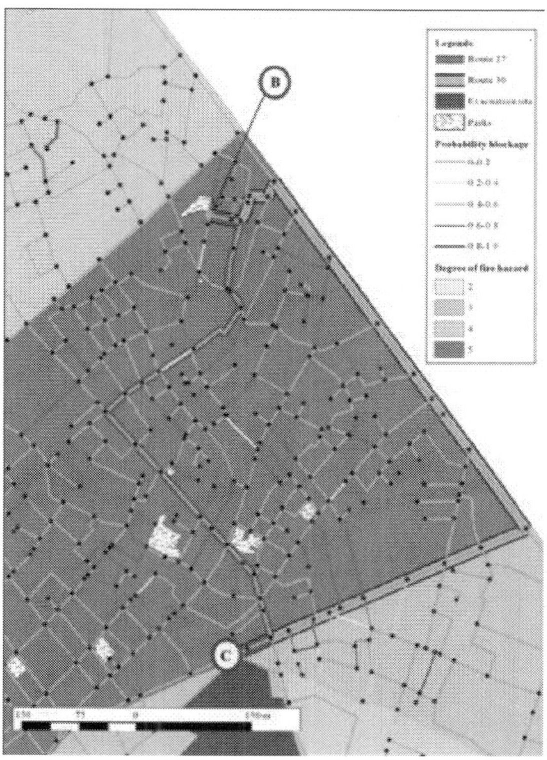

Figure 16: Preference routes from temporary gathering place to large-area evacuation site.

Figure 17: Preferences routes from temporary gathering place to large-area evacuation site, shown with the estimated evacuation populations in roads and for each residence.

Figure 18: Preferences routes from temporary gathering place to large-area evacuation site, shown with road widths.

Concerning the derived routes, the maximum evacuation distance, which belongs to Route 32, is only 1106.7 m. Therefore, even if the length of either of the routes in Case (1) described above (routes from Point A, the evacuation starting point, to the neighborhood of the temporary gathering place) is added to the length of any of the derived routes in Table 6, the combined distance will still fall within two kilometers. Similarly, concerning evacuation time, even if the evacuation time of either of the routes in Case (1) is added to that of any of the derived routes in Table 6, the combined evacuation time will still be within an hour. Concerning arrival probability, routes which fulfilled the preference condition of greater than or equal to 70% were Routes 26, 27, 30 and 31. Of these four routes, Routes 27 and 30 (Figure 16), which have a relatively low degree of hazard due to fire in the road and whose characteristics differ from each other, were chosen as preference solutions. Route 27 has a short evacuation distance, and also, compared to the other derived routes, a relatively high arrival probability. However, from Figure 17 and Figure 18, it can be seen that for this route, the distance passing through roads in blocks/chome with a narrow width is long, and the degree of hazard due to fire is relatively high. Route 30 has a relatively long evacuation distance; however, it has a high arrival probability and also a low degree of hazard due to fire. From Figure 16 it can be seen that roads in this route are unlikely to be crowded, because this route avoids roads with a high blockage probability, and evacuees who take this route will come out at a road with a wide width after going only a short distance; further, the route passes through roads with wide widths. Therefore, this route has a shorter evacuation time than Route 27, which has a short route distance.

Evacuation Routes in the Case of Direct Evacuation

Twenty-one routes were derived as Pareto optimal solutions for direct evacuation routes from Point A, the evacuation starting point, to Point C, near the large-area evacuation site. Table 7 shows each of the evaluation values for the derived routes, and Routes 19 and 20 were selected for illustration in Figure 19. Further, Figure 20 and Figure 21 show results for preference routes superposed with the estimated evacuation populations in roads and for each residence, and derived routes superposed with road width, respectively.

Table 7: Evaluation values for routes in case of direct evacuation

Derived route	Evacuation distance [m]	Arrival probability	Degree of hazard due to fire in the road	Evacuation time [h]
1	543.5	0.0052	1839.5	0.43
2	559.9	0.0110	1921.3	0.46
3	561.8	0.0020	1794.9	0.45
4	583.8	0.0190	2040.8	0.49

5	598.4	0.1100	2113.6	0.56
6	600.0	0.0170	1525.1	0.47
7	600.2	0.0430	2122.6	0.53
8	608.8	0.0320	1569.4	0.51
9	614.7	0.2400	2195.4	0.61
10	616.6	0.0430	2069.1	0.60
11	619.4	0.0720	2218.8	0.39
12	623.2	0.0350	1413.3	0.48
13	625.4	0.0130	1424.2	0.47
14	629.6	0.0350	1171.3	0.44
15	632.0	0.0640	1457.6	0.51
16	638.5	0.0640	1215.5	0.47
17	654.8	0.3600	1799.3	0.62
18	678.0	0.7400	1687.5	0.60
19	684.5	0.7400	1445.4	0.56
20	975.4	1.0000	1360.1	0.65
21	1074.9	0.4900	1018.3	0.72

The derived routes all had an evacuation distance of within two kilometers and an evacuation time of within one hour. Of the derived routes, only three had an arrival probability of over 70%—Routes 18, 19 and 20. Near the evacuation starting point, there are many roads with a high road blockage probability, so it is considered that evacuation is difficult. However, it was possible to find routes such as these three—routes with a high arrival probability and a low fire hazard. Of these three routes, Routes 19 and 20 (Figure 19) have a relatively low risk due to fire compared to the other derived routes, because they enable speedy evacuation to roads with a wide width. These routes were selected as preference solutions. Further, Routes 19 and 20 also have a high arrival probability; therefore, fast and safe evacuation is possible along these routes.

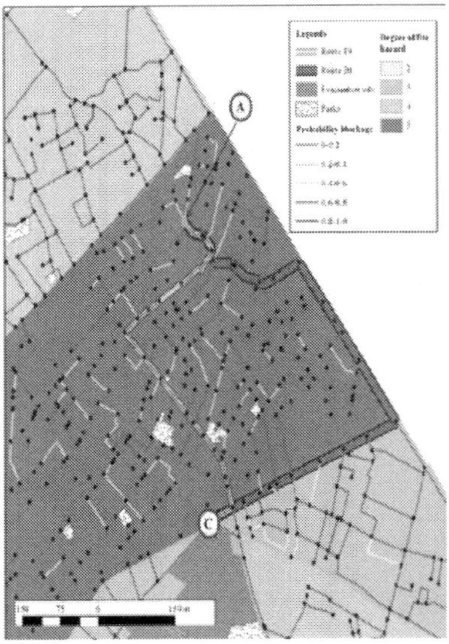

Figure 19: Preference routes for case of direct evacuation.

Figure 20: Preference routes for case of direct evacuation, shown with the estimated evacuation populations in roads and for each residence.

Figure 21: Preference routes for case of direct evacuation, shown with road width.

CONCLUSIONS AND FUTURE RESEARCH TOPICS

The conclusion of this study can be summarized into the following three points:

- A GA was used to design and create an evacuation route search algorithm which solves the problem of the optimization of earthquake disaster evacuation routes by treating it as an optimization problem with multiple objectives, such as evacuation distance and evacuation time. By using this method, it is possible to find multiple evacuation route candidates whose objectives are in a trade-off relationship with each other in a single search. Further, selecting preference solutions from among the evacuation route candidates derived enables the determination of multiple evacuation routes in preparation for disaster.

- In this method, goodness of fit is set by using a Pareto ranking method to determine the ranking of individuals based on their relative superiorities and inferiorities. This enables objective functions to be set without taking into account weightings or orders of priority between each objective. Therefore, the number of objectives can be freely increased, and various objectives which disaster evacuation routes are required to meet can be taken into account. Further, because this method performs searches based on road blockage probability and evacuation distance, which are quantified data, derived routes are routes that have been quantitatively evaluated, and they can be highly effective indicators for determining evacuation routes.
- In the method, road information concerning obstacles to disaster evacuation was created based on data regarding present conditions, such as building and land use data. In this method, searching for evacuation routes based on the information on present conditions allows evacuation routes to be derived based on present building and road locations. Therefore, this method has a high degree of spatial reproducibility. Further, this method is based on publicly available information; therefore, obtaining geographic information similar to that of this study enables this method to be effective regardless of what region it is applied to, or whether the data regards the past or the future. Therefore, in addition to spatial reproducibility, this method also has a high degree of temporal reproducibility.

Examples of future topics for research are improvement of search accuracy, GIS visualization of trade-off relationships of derived solutions, and further refinement of road information.

REFERENCES

1. The Reconstruction Design Council in Response to the Great East Japan Earthquake (2011) Towards Reconstruction. "Hope beyond the Disaster". Bureau of Task Force for Reconstruction in Response to the Great East Japan Earthquake, 84.

2. Komura, T. and Hirano, A. (1997) Disaster Imagination Game (DIG), A Drill Using Maps. The Annual Conference of the Institute of Social Safety Science, 7, 136-139.
3. Liu, Y., Hatayama, M. and Okada, N. (2006) Development of an Adaptive Evacuation Route Algorithm under Flood Disaster. Annuals of the Disaster Prevention Research Institute, 49B, 189-195.
4. Takeuchi, T. and Kondou, A. (2002) An Analysis of Evacuation Routes Taking into Consideration Road Sections Blocked by Earthquake—A Case Study of Sazaki City. The 26th Infrastructure Planning and Management Research Presentation Meeting, Presentation No. 332(CD-ROM).
5. Ichikawa, F., Sakata, T. and Yoshikawa, T. (2001) Study on the Placement of Evacuation Areas, Taking into Account the Danger of Blocked Roads on Evacuation Routes. Summaries of Technical Papers of Annual Meeting Architectural Institute of Japan, 489-490.
6. Inoue, M. and Yamamoto, K. (2013) Method for Evaluating the Location of Tourist-Related Public Facilities Using Genetic Algorithms and GIS. Journal of Communication and Computer, 10, 496-512.
7. Asou, T., Matsumoto, Y. and Morishita, K. (2007) A Study on the Optimal Location of Disaster Shelters Using Genetic Algorithm. Research Report of Faculty of Engineering, 58, 522-533.
8. Koarai, M., Satou, H. and Une, H. (2007) GIS Analysis of the Relationship between Earthquake-Induced Ground Disaster and the Physical Characteristics of Land. Bulletin of the Geospatial Information Authority of Japan, 112, 115- 123.
9. Hara, K., Tsukahara, S., Kanou, H., Sota, T. and Kurokawa, H. (2006) Predictive Route Searching for Car Navigation Systems Using Multi-Objective Genetic Algorithm. Society for the Study of Intelligent Transport Systems, Information Processing Society of Japan, ITS-24, 31-38.
10. Kinoshita, Y. and Yokokishizawa, H. (2008) Proposal for a Tourist Route Planning Support System Using Sensitivity Evaluation and Multi-Objective GAs. Japan Society for Fuzzy Theory and Intelligent Informatics, 24th Fuzzy System Symposium, FC2-1, 864-869.

11. Bureau of Urban Development, Tokyo Metropolitan Government (2013) District-Based Assessment of Vulnerability to Earthquake Disaster (No. 7). 300.
12. Urban Disaster Prevention Office of the City Bureau of the Ministry of Construction (1997) Practical Handbook of Urban Disaster Prevention. Earthquake Disaster Prevention, 249.
13. Director-General for Disaster Management in the Cabinet Office (2005) Report on Investigation into Wide-Area Disaster Prevention Measures in the "S.K.Y Kouikiken"* Area (*ALarge Area Centering on the Fuji, Hakone and Izu Areas). 167.
14. Tokyo Metropolitan Government. Disaster Prevention Facilities, Sheltersand Evacuation Areas. http://www.bousai.metro.tokyo.jp/bousai/1000029/index.html
15. Sumida City. Sumida City Disaster Prevention Map.http://www.city.sumida.lg.jp/kurasi_guide/katei_tiikinobousai/sumidakubousaitizu.html
16. Sumida City. Your Evacuation Site.http://www.city.sumida.lg.jp/kurasi_guide/katei_tiikinobousai/hinanbasyo.html
17. Yoshikawa, T., Yokota, K. and Sakata, T. (2003) A Stochastic Model Basis for Simulation of Road Blockage Which Takes into Account Individual Collapsed Buildings in Actual Urban Areas in Great Earthquakes. Summaries of Technical Papers of Annual Meeting Architectural Institute of Japan, 991-992.

Chapter 10

Potential Hazard Map for Snow Disaster Prevention using GIS-Based Weighted Linear Combination Analysis and Remote Sensing Techniques: A Case Study in Northern Xinjiang, China

Gulijianati Abake[1], Ahmad Al-Hanbali[2], Bayan Alsaa-ideh[2], and Ryutaro Tateishi[2]

[1]Graduate School of Science, Chiba University, Chiba, Japan

[2]Center for Environmental Remote Sensing, Chiba University, Chiba, Japan

ABSTRACT

Snow disaster is one of the top ten natural disasters worldwide. Almost every year, there will be snow disasters in north Xinjiang, northwestern China. Since the accumulated heavy snow in winter season will seriously threaten people's lives, the main object of this study is to produce a potential hazard map for snow avalanche prevention. Taking three snow seasons from November to March of year 2008 to 2010, potential hazard areas were estimated, based on snow volume products and terrain features. Remote sensing (RS) techniques and geographical information system (GIS) based weighted linear combination (WLC) approach were applied, taking into consideration multiple criteria. Snow avalanche risks were analyzed using physical exposure and vulnerability indexes. The analysis indicates that: the areas at high-risk of avalanches are located in the north and south part of the counties of Altay, Bortala and Ili prefectures; the areas at medium-risk of avalanches are found in the certain part of Altay prefecture and Urumqi, Changji, Tacheng prefectures; the avalanche risk is generally low throughout the large area to the certain part of the study area and the region on the border of the eastern north Xinjiang. Overall, the risks of snow avalanche in Altay and Ili prefectures are higher than that other regions; those areas should be allocated correspondingly more salvage materials.

INTRODUCTION

Snow avalanches (hereafter also called avalanches) are a natural disaster that has considerable influence on human activities in mountain regions. Snow avalanches are rapid mass movements occurring in snow covered mountain areas all over the world [1]. Almost every year, there will be snow disaster in Xinjiang [2]. Snow avalanche is one of the serious disasters in north Xinjiang, China, with the characteristics of sudden, fast-movement, large destructive and so on. The massive snow accumulation frequently causes disasters such as livestock casualties, traffic jams and other disasters, causing considerable economic loss and serious damage to towns and farms and leading to local people's production and life affected. Over half of precipitation in the study area falls as snow in the cold season (October to April), and over 84%

of precipitation concentrates in the mountains [3] . Therefore, the research on snow avalanche is very critical to general public for snow disaster prevention.

In order to effectively conduct the investigation on this district, it has these conditions: 1) very wide area with a total land area of 39,845,600 ha; 2) areas greatly different in elevation from 1000 m to 6300 and 3) serve climate conditions depending on the season to reject human access for field survey. Conventional station observation can hardly meet practical demands. Remote sensing (RS) techniques and geographical information system (GIS) can largely improve the effectiveness and efficiency of snow avalanche monitoring. RS technology has advantages such as large-scale, rapid, cyclical, multi-scale, multi-temporal, etc. GIS has the capability to handle the necessary data gathered from various sources. It combines spatial data with quantitate, qualitative, and descriptive information database, which can support a wide range of spatial queries. Many researchers study and investigate the snow avalanche hazard. However, previous studies almost concentrate on Tien Shan Mountains regions. In this study, we focus on whole north Xinjiang area, including Tien Shan Mountains region and Altay Mountains region. Because north Xinjiang, its land area, consists of two large mountain ranges with the basin in between. The problem of living with this natural hazard is known and must be addressed. Moreover, it requires a more detailed description of the avalanche hazard area. Based on comprehensive analysis over current research status at home and abroad and previous studies, this paper aims to produce a potential hazard map for snow avalanche prevention on north Xinjiang, China. RS techniques and a GIS-based weighted linear combination method are put into the research on the establishment of potential avalanche hazard map, on the basis of rules related to topographical characteristics. Avalanche risks were analyzed with physical exposure and vulnerability indexes overlay to the potential hazard map, on the basis of available information.

STUDY AREA AND DATA USED

Study Area

The northern part of Xinjiang Uyghur Autonomous Region (hereafter called north Xinjiang) located on north- western China was selected as a case study area (Figure 1). The extreme minimum temperature may reach −40°C and the mean duration of snow cover is about 130 days per year, generally from the middle of November through March. In particular, northern Xinjiang is one of three snow centers in China. It has a relatively rich seasonal snow source due to winter circulation around the Siberian High and southern mountain blockage. Over half of precipitation in the study area falls as snow in the cold season (November to April), and over 84% of precipitation concentrates in the mountains [4] . Snow avalanches which are the major threat to lives and property every winter season such as livestock casualties, traffic jams and other disasters over the study area cause considerable economic loss and serious damage to towns and farms.

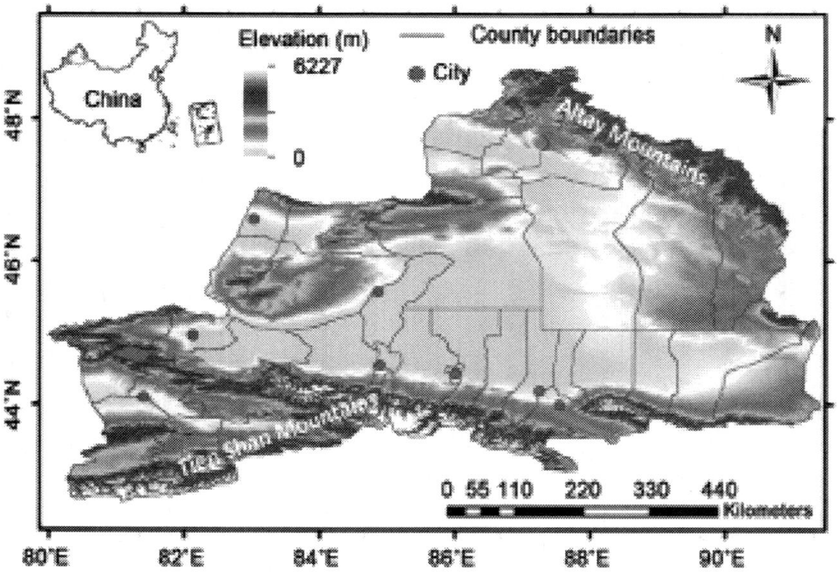

Figure 1: The case study area.

Data Used

Available remote sensing products exist at a wide variety of spatial and temporal resolutions. The data set used in this research is: 1) The Moderate Resolution Imaging Spectra radiometer (MODIS) Snow Cover 8-day L3 Global 500 m Grid product. This product has been provided through the Distributed Active Archive Center (DAAC) of the National Snow and Ice Data Center (NSIDC). The data set contains data fields for maximum snow cover extent over an eight-day compositing period and a chronology of snow occurrence observations. The data are based on a snow mapping algorithm that utilize a Normalized Difference Snow Index (NDSI) and two additional criteria tests: the thresholds of Band 2 and Band 4 (http://nsidc.org/data/modis/data_summaries/index.html). 2) The Advanced Microwave Scanning Radiometer-Earth Observing System (AMSR-E) monthly L3 Global Snow Water Equivalent (SWE) product. This product has been available from the Distributed Active Archive Center (DAAC) of the National Snow and Ice Data Center (NSIDC) since June 2002. This product is generated from AMSR-E/Aqua L2A Global Swath Spatially Resampled Brightness Temperatures data. For each low frequency (<89 GHz) sample, snow depth retrieval is performed through snow maps masking and brightness temperatures classifying. And then retrievals and calibrated and projected into 25 km. Equal-Area Scalable Earth Grids (EASE-Grids) (http://nsidc.org/data/amsre/data_summaries/index.html). 3) Canadian Meteorological Centre (CMC) Daily Snow Depth Analysis Data. The data set consists of Northern Hemisphere snow depth analysis data processed by the Canadian Meteorological Center (CMC). Snow depth data obtained from surface sy- noptic observation (synops), meteorological aviation reports (metars), and special aviation reports (SAs) were acquired from the World Meteorological Organization (WMO) information system for use in the CMC analyses (http://nsidc.org/api/metadata?id=nsidc-0447). 4) The Advanced Space borne Thermal Emission and Reflection Radiometer (ASTER) Global Digital Elevation Model (GDEM) version 2; this product was produced in cooperation with the Japan-US ASTER Science Team using ASTER data (http://asterweb.jpl.nasa.gov/gdem.asp). 5) Statistical data. These profile data obtained from Xinjiang Statistical Yearbook were published by China Statistics Press.

The characteristics of those data used in this research are summarized in Table 1.

Table 1: The characteristics of data used in this study

Data	Spatial resolution	Tile	Source
MODIS	500 m	h23v04, h24v04	NSIDC
AMSR-E	25 km		NSIDC
Snow depth	24 km		CMC
GDEM	30 m	N30E70 - N50E100	USGS
Statisctical data			

METHODOLOGY

The first step is building a hierarchy which can model the problem. Snow volume, altitude, and slope selected as an evaluation factor for potential avalanche danger model, since snow accumulation and steep topography are prerequisites of avalanches. In this study, snow avalanche danger is the decision goal, and snow volume, elevation, and slope are selected as fundamental evaluation factors in the model. GIS has the capability to handle and simulate the necessary data gathered from various sources [5] . GIS combines spatial data with quantitative, qualitative, and descriptive information database, which can support a wide range of spatial queries. All of these factors have made GIS an essential tool for potential hazard area estimation. A multi-criteria evaluation (MCE) method can serve to inventories, classify, analyze and conveniently arrange the available information concerning choice-possibilities in regional planning. It is mainly involved with how to combine the information from several criteria to form a single complex of evaluation. It is used to deal with difficulties that decision makers encounter in handling large amounts of complex information. Weighted linear combination (WLC) is one of the widely used MCE methods for land suitability analysis. WLC is a concept which combines maps by applying a standardized score to each class of a certain parameter and a factor weight to the parameters themselves. It involves standardization of the suitability maps, assigning the weights of relative importance to the suitability's maps, and then combining the weights and standardized suitability maps to obtain an overall suitability score. The WLC analysis was applied using the following equation:

$$H = \sum W_i \times X_i \tag{1}$$

Where H is the Dangerous degree, W_i is a weighting of factor i, and X_i is the criteria score of factor i. In this research, using a GIS-based WLC analysis, examines a major three possible factors for a snow avalanche problem, taking into consideration multiple criteria. In order to use GIS for decide potential snow avalanche hazard area, data were stored in the GIS system and each parameter is considered as layer. To apply the WLC analysis practically, ArcGIS software package were used. ArcGIS software uses a weighted sum analysis that is act as a WLC analysis. A weighted sum analysis provides the ability to weight and combine multiple inputs to create an integrated analysis. In other words, it combines multiple raster inputs, representing multiple factors, of different weights or relative importance. It is one of common methodologies used for snow disaster monitoring in general, and for defining potential hazard area in particular.

Produce Snow Volume Product

Snow avalanche is a falling mass of snow. A large volume of snow is necessary for an avalanche to occur. Snow volume is sum of all non-zero snow depth pixels multiplied by the snow covered area pixel [6] . The biggest problem in analyzing a snow-cover time series is cloud coverage. In practice, it was difficult to find a MODIS image that was no cloud contamination for the Xinjiang area [7] . Clear-sky conditions are rare, especially in the winter season [8] .

Figure 2 displays the monthly cloud-cover percentage for four selected years in northern Xinjiang. The maximum cloud coverage is mainly concentrated during winter months (The percentage of cloud coverage mainly concentrated on winter season). The overall mean cloud-cover percentage in winter months varies from 20% for 2008, down to 10% for 2009, increase to 20% for 2010, and over 22%

for 2011. It significantly limits the ability to map the exact number of snow-covered days. However, when aiming at snow-cover extent analyses over time, a cloud-free time series is a major prerequisite.

Generation of snow volume product mainly consist of following two steps: first, produce cloud-free snow- covered product; then, produce snow volume product. The flowchart for the produce snow volume product is shown in Figure 3. Optical sensors such as MODIS have high classification accuracy in clear-sky condition while it is much lower if cloud is counted. Space-borne passive microwave data, such as the AMSR-E data, has the advantage of all-weather imaging but a much coarse resolution. The algorithm which for merging two re- mote sensing products (utilizing MODIS snow cover products and AMSR-E SWE products) with complementary spatiotemporal resolution characteristics is taken to generate cloud-free snow cover products, similar to Gao et al. (2010) [9] . A method consists of unifying codes, combing products. 1) Unifying codes: The original integer numbers in the MODIS and AMSR-E products were then transformed to new codes and used for the combination processes. 2) Combining products: replace the remaining cloud or other contamination of MODIS using the value of AMSR-E Snow Cover image. According to the definition of snow volume, snow volume product generated from multiplication of cloud-free snow-covered area with snow depth.

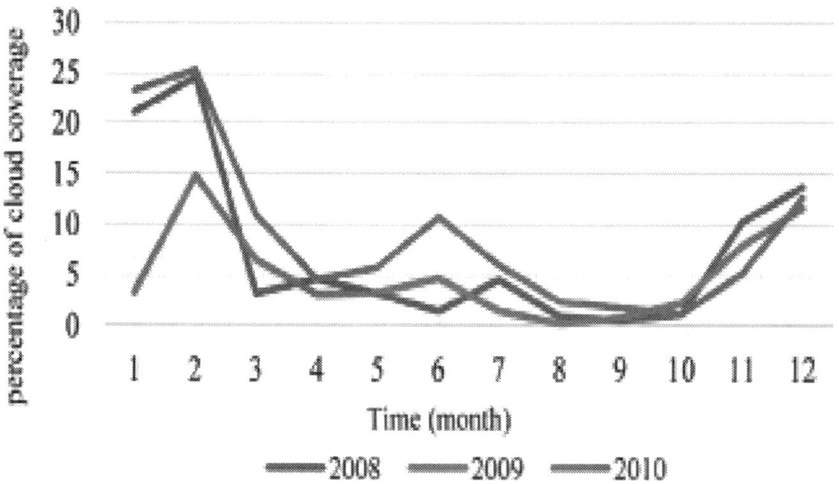

Figure 2: The state of cloud coverage in north Xinjiang.

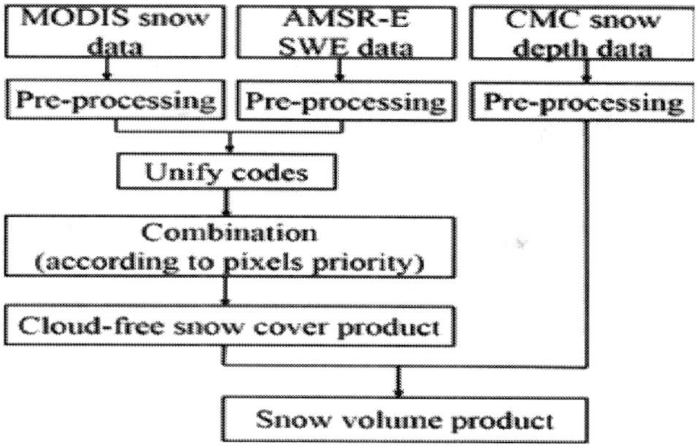

Figure 3: Flowchart for produce snow volume product.

Define Potential Hazard Area

Potential hazard map is produced through spatial data process operation (Figure 4). For each evaluation factor, reclassification system is required. Moreover, overly calculation over all of evaluation factors' data need to be executed so that the score of each study unit's snow avalanche danger could be calculated and then danger grade could be made according to operation system. Detail explanation about the step that how to perform weighted sum analysis are:

- All three input data were given scores to reclassify. Since these three input layers are in different numbering systems and with different ranges, raster's of these data could not be simply added. For each factor, reclassification system is required. The score assess their importance on snow avalanche occurrence on study area that range from 0 to 10 (Table 2). The higher score was obtained, the more importance it would present. A score of 0 indicates no potential danger, and a score of 10 indicates an extremely serious dangers. The total values with the higher scores will represent the events most in need of organization planning for emergency preparedness. In this study, the criteria for slope and altitude in selected parameters based on the local condition of study area, the criteria for slope from the method of [10] . They were proceeded map overlaying analyses in GIS for the values of grids. This method can provide the decision makers several options for selecting appropriate strategies, since using this method, the final output map will range from the "high dangerous" to "no dangerous".

- Identify weights for these three input data. Evaluation factors in the model may have different importance. Before the factors are combined the factors can be weighted, based on their importance (Table 2). Weights were generally assigned to these maps to express the relative importance. The total weight should be added up to 100% in order for the output map to be meaningful and consistent, and the attribute score must be chosen using a scheme that was the same for each map. As a general rule, it was decided to give higher weightings to factors that affect directly on the community. In this study, the three selected evaluation factors are considered as almost equal important in pair-wise comparison.

Then, a final composite map was produced using WLC. The cell values of each input raster are multiplied by the raster's weight. The resulting cell value is added to produce the final output raster. Based on current understanding for study area's snow avalanche occurrence mechanism, the layers, the criteria, the reclassification system and their weights are listed (summarized) in Table 2.

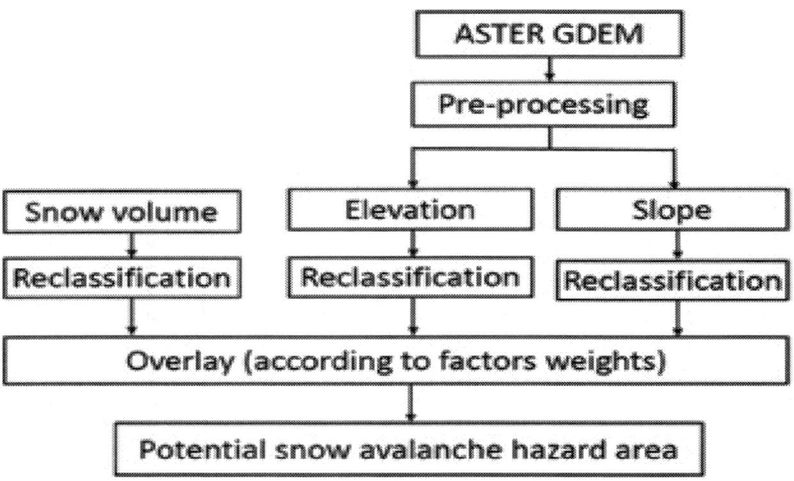

Figure 4: Spatial data process operation for define potential hazard area in north Xinjiang.

Table 2: Attribute scores and weights for the map used in the avalanche hazard decision

Category	Spatial resolution	Score	Weights
Snow volume	2500 < V	1	0.4
	2500 < V < 5000	4	
	5000 < V < 7500	7	
	7500 < V	10	
Altitude	1 < H < 1000	4	0.3
	1000 < H < 3500	7	
	3500 < H	10	
Slope	< 30	4	0.3
	30 < < 40	7	
	40 <	10	

EXPERIMENTAL RESULTS AND DISCUSSION

Comparison of MODIS Snow Products with Cloud-Free Snow Cover Products

Figure 5 shows comparison between MODIS snow products and cloud-free snow cover products on monthly time scales. X axis indicate time (month) scale, Y axis indicate percentage of each factor. By taking advantage of both high spatial resolution of optical data and cloud transparency of passive microwave data, the combination of MODIS and AMSR-E removed all the cloud and other contaminations, greatly complements the deficiency of MODIS product. This process proves that produce cloud-free snow cover product by merging existing MODIS and AMSRE Snow product is practicable. The results shows that the new monthly cloud-free snow cover product at 500 m resolution can effectively capture snow covered information during January 2008 to De- cember 2010.

Dynamics of Snow Volume

Outputs presents that the total snow volume of 2008 and 2009 is lower than the total snow volume of 2010 (Figure 6). The time variation of the snow volume is almost dominated by the seasonal cycle. Results show that the mean annual snow volume over the whole study is domain equal to (48) km^3, (60) km^3, and (89) km^3 over this three year time period, respectively. The mean monthly snow volume ranges from about 38 × 10^9 m^3 in summer to about 104 × 10^9 m^3 in winter. The serious snow volume from December to March, accounted for 80% of the total snow volume. From November to March, snow volume was increasing (this season called snow accumulation season), this heavy snow accumulation leads to snow disasters, such as snow avalanches, snow drifting, etc. From March to June, snow volume was decreasing (this season called snow melting season), this type serious snow melting may occurred floods, name as snow-melt flood. In addition, serious snow volume change mainly concentrated on Altay, Tacheng, and Ili regions. The monthly variation of snow volume time series over the study area during January 2008 and December 2010 are shown Figures 7-9.

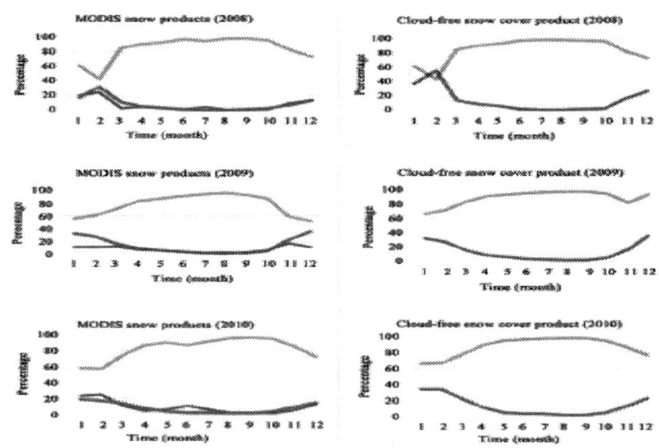

Figure 5: Comparison of MODIS snow products with cloud-free snow cover products (red line shows snow; green line shows non-snow; grey line shows cloud).

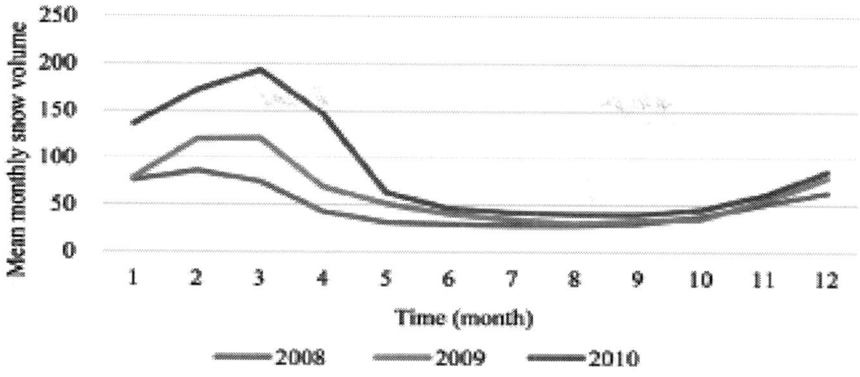

Figure 6: The state of mean monthly snow volume variation.

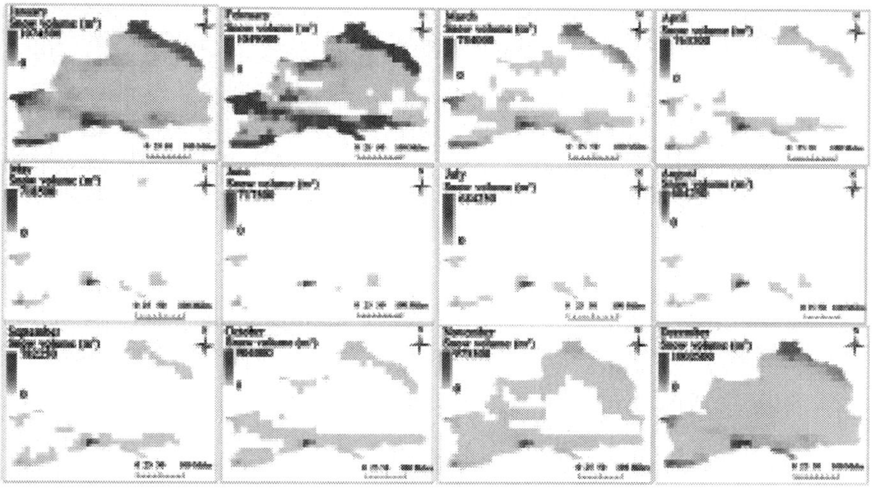

Figure 7: The mean monthly snow volume variation (2008).

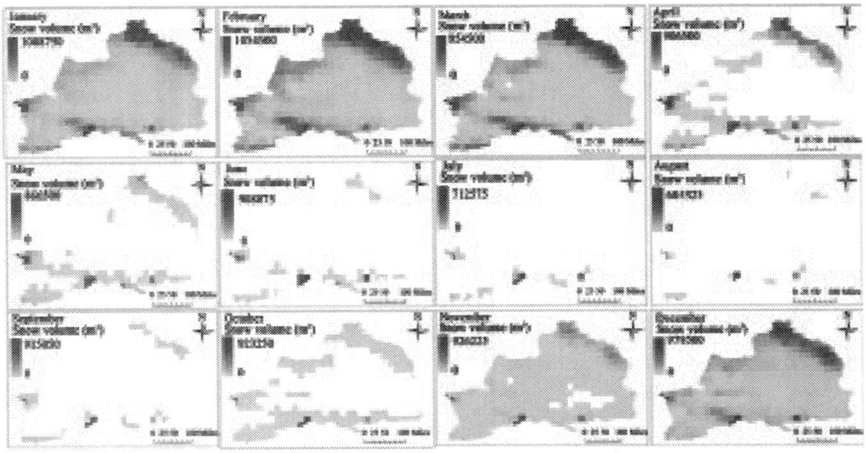

Figure 8: The mean monthly snow volume variation (2009).

From November to March, snow volume was increasing (this season called snow accumulation season), this heavy snow accumulation leads to snow disasters, such as snow avalanches, snow drifting, etc. From March to June, snow volume was decreasing (this season called snow melting season), this type serious snow melting may occurred floods, name as snow-melt flood.

Potential Avalanche Hazard Map

The resulting potential hazard map for snow avalanche prevention was shown in Figure 10. The dangerous grade for potential hazard map were named: no danger (zone I), very low danger (zone II), moderate danger (zone III), high danger (zone IV), and very high danger (zone V). As shown in Figure 10, the areas threatened by very high and high of potential snow avalanche hazard are located in the five counties of Habahe (Kaba), Bur- qin, Zhaosu (Mongolkure), Tekas, and Nilka, which are located in the north section of Altay mountains and in the south and east section of Tien Shan Mountains. In these areas, the number of days with snow cover every year can exceed 200, with the longest snowfall period, beginning earlier and ending later than in other regions. During the period from November to April, the areas covered with permanent snow are more susceptible to snow disaster.

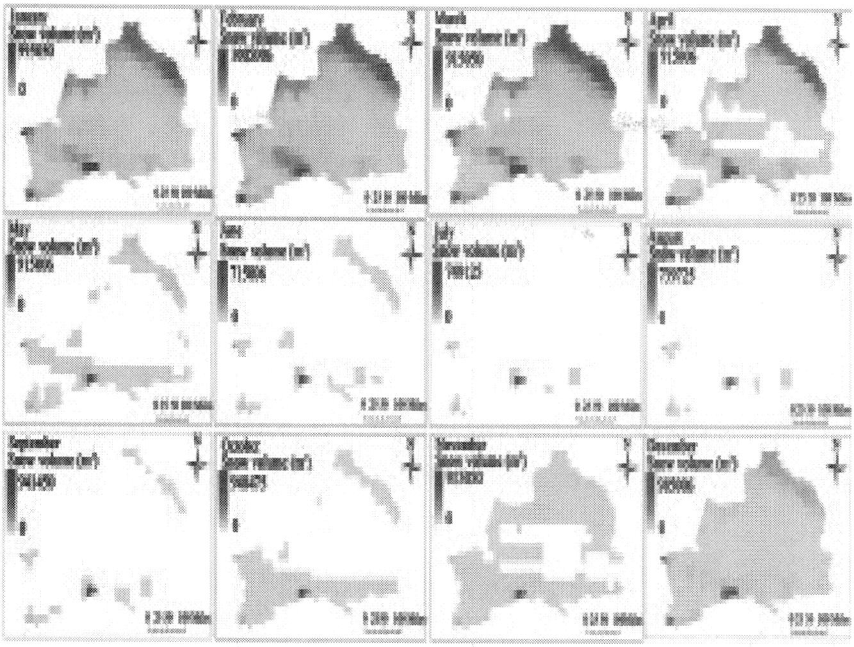

Figure 9: The mean monthly snow volume variation (2010).

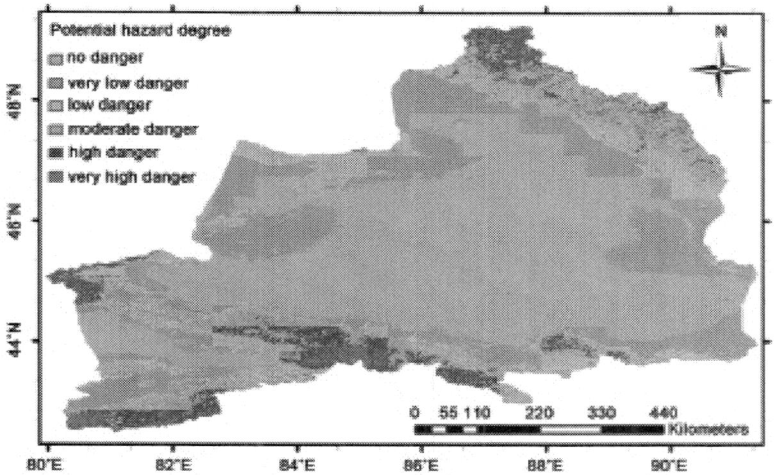

Figure 10: Potential hazard map for snow avalanche prevention in north Xinjiang.

Areas at moderate danger of potential snow avalanche hazard are found the following 13 counties: Altay, Fuhai, Fuyun, Qinghe, Wenquan, Hupcheng, Xinyuan, Usu, Shawan, Changji, Fukang, Jimisar, Hutubi, which are scattered across the Altay prefecture in the north, Tacheng, Bortala and Ili prefectures in west and Changji prefecture in the south. These areas cover at elevations from 1000 m to 3500 m. The danger of potential snow avalanche hazard is generally low throughout the area in Qitai, MoriKazak, Manas, Xinyuan, Qapqal, Bole, Jinghe, Yumin, Tuoli, Tacheng, Emin, and jeminay, which are located on the east part of Altay prefecture, middle of Ili prefecture, and Tacheng prefecture. The dangerous free area of potential hazard includes Junggar basin and its surroundings. Overall, the relative high dangerous areas of potential snow avalanche hazard area founded in Ili prefecture in the south, and Altay prefecture in the north than other regions.

RISK ANALYSIS IN NORTH XINJIANG

In the 1990s, scholars recognized that natural disasters had both natural and social attributes [11] . As mentioned in the previous section, the number of natural disasters has been increasing, as has their impact, due to such external changes as the concentration of populations and property in hazardous areas. Snow avalanche, torrential rains, earthquakes, are natural phenomena we refer to as "hazards" and are not considered to be disasters in and of themselves. For instance, a snow avalanche that occurs on a top of alpine does not trigger a disaster because there is no existing population or property affected. In addition to a hazard, some "vulnerability" to the natural phenomenon must be present for an event to constitute a natural disaster. "Vulnerability" is defined as a condition resulting from physical, social, economic, and environmental factors or processes, which increases the susceptibility of a community to the impact of a hazard. "Exposure" is another component of disaster risk, and refers to that which is affected by natural disasters, such as people and property. In general, "risk" is defined as the expectation value of losses (deaths, injuries, property, etc.) that would be caused by a hazard. Disaster risk can be seen as a function of the hazard, exposure and vulnerability as follows:

$$\text{Disaster Risk} = \text{function}(\text{Hazard, Exposure, Vulnerability}) \quad (2)$$

Growing exposure and delays in reducing vulnerabilities result in an increased number of natural disasters and greater levels of loss. In present paper, snow avalanche risk were analyzed using physical exposure and vulnerability indexes, which including population density, livestock number at the end of year, and highway network (Figure 11). They reflect the risk assessment of snow avalanches on the north Xinjiang.

The snow avalanche risk analysis revealed that: the areas at high-risk of snow avalanche on the north Xinjiang area are mainly located in north and south part of the counties of Altay, Bortala and Ili prefectures. These areas influenced by northern cold air. With the longest snow fall period, in these areas, the number of snow covered days can exceed 200, beginning earlier and ending later than in other region. During the period from November to April, these areas covered with permanent snow are more susceptible to snow avalanches. The areas at medium-risk of snow avalanches are found in certain part of Altay prefecture, Urumqi, Changji, and Tacheng prefectures. These areas cover at elevation from 1000 m to 3500 m. The risk of snow avalanches is generally low throughout the large area to the certain part of study area and the region on the border of the eastern north Xinjiang. Overall, the avalanche risk of snow disaster in Altay and Ili prefectures are higher than that other regions. This study suggests that those areas should be allocated correspondingly more salvage materials. To reduce snow disaster risk, it is important to reduce the level of vulnerability and to keep exposure as far away from hazards as possible by relocating populations and property. The reduction of vulnerability can be achieved through such measures as mitigation and preparedness.

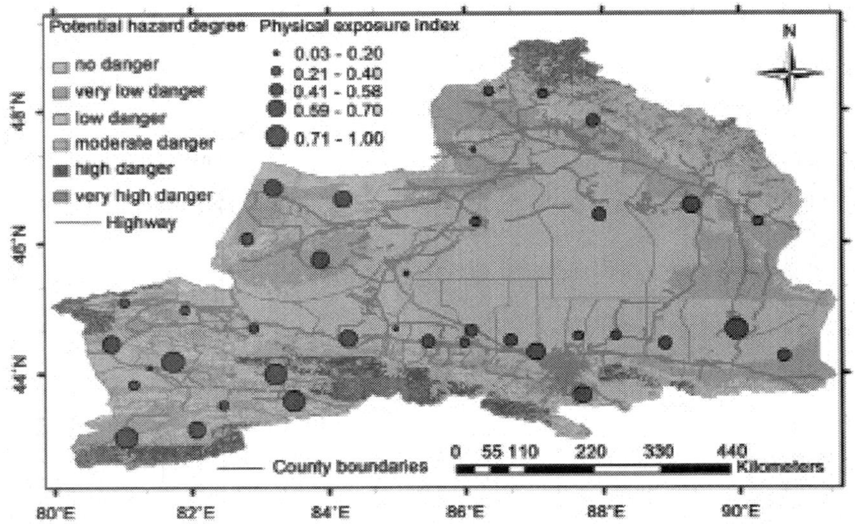

Figure 11: Snow avalanche analysis for north Xinjiang area.

SUMMARY

The north Xinjiang is a region of China that frequently experiences snow avalanche disaster, which affects people's lives and productivity, causing serious damage to towns and farms. An accurate definition of the potential hazard area and the assessment of snow avalanche risk are important if the hazard of snow disaster is to be mitigated. With the help of avalanche theories formulated in China and abroad, the final target of this study has been achieved in terms of the potential hazard map for snow avalanche prevention. The theoretical framework of snow avalanche disaster risk is described, which has a guide meaning in snow avalanche disaster mitigation. An index of physical exposure and vulnerability, including population density and livestock number at the end of year, highway network, is established based on the data from Xinjiang statistical yearbook. They reflect the risk assessment of snow avalanches on the north Xinjiang. The need has been established for better analysis, communication, and visualization of avalanche hazard to aid decision-making. The conclusions of this study are as follows: 1) Using RS techniques and GIS-based weighted linear combination method to scientifically map potential avalanche dangerous area for snow avalanche prevention, based on snow volume product and terrain features, over 2008 to 2010 for north Xinjiang area and visualize the results. According to the results, we could obtain information for disaster management. For example, we could obtain the danger level of each region. The methods have certain advantages, such as having simple processes and providing quantitative, comparative, synthetic, analytic results. 2) Snow volume products were calculated from generated cloud-free snow covered product and existing snow depth data. In order to obtain cloud free snow covered data, MODIS snow covered data and AMSR-E data were merged, since, in practice it was difficult to find a satellite image that was no cloud contamination for the Xinjiang area. Positive combination of both MODIS snow covered data and AMSR-E SWE data, removed the all cloud area of MODIS data. Snow volume product will be the first snow volume product in the north Xinjiang area, and this study provides accurate snow volume information temporally and spatially. 3) The snow avalanche risk analysis revealed that: the areas at high-risk of snow avalanche on the north Xinjiang area mainly located in the north and south part of the counties of Altay, Bortala and Ili prefectures. These areas are influenced by the northern cold air. With the longest snow fall period, in these areas, the number of snow covered days can exceed 200, beginning earlier and ending later than in other regions. During the period from November to April, these areas covered with permanent snow are more susceptible to snow avalanches. The areas at medium-risk of snow avalanches are found in the certain part of Altay prefecture, Urumqi, Changji, and Tacheng prefectures. In this area, coverages are at elevation from 1000 m to 3500 m. The risk of snow avalanches is generally low throughout the large area to the certain part of the study area and the region on the border of the eastern north Xinjiang. Overall, the avalanche risk of snow disaster in Altay and Ili prefectures is higher than the other regions. This study suggests that those areas should be correspondingly allocated more salvage materials. The potential hazard map and avalanche risk analysis in this study are specifically intended to support the local government and native people in the future, and hope to play a role in the prevention of snow avalanches. Future effort is required to more advanced data acquisition techniques which should be introduced into the monitoring model.

ACKNOWLEDGEMENTS

The authors would like to appreciate the Distributed Active Archive Center (DAAC) of National Snow and Ice Data Center (NSIDC), Canadian Meteorological Centre (CMC), and US Geological Survey (USGS) for providing the free products to support kinds of remote sensing researches.

REFERENCES

1. Eckerstorfer, M. (2012) Snow Avalanche in Central Svalbard: A Field Study of Meteorological and Topographical Triggering Factors and Geomorphological Significance. Ph.D. Thesis, University of Oslo, Oslo.
2. Su, W., Zhang, X., Wang, Z., Su, X., Huang, J., Yang, S. and Liu, S. (2011) Analyzing Disaster Forming Environments and the Spatial Distribution of Flood Disasters and Snow Disasters that Occurred in China from 1949 to 2000. Mathe- matical and Computer Modelling, 54, 1069-1078. http://dx.doi.org/10.1016/j.mcm.2010.11.037
3. Sun, G., Chen, Y., Li, W., Pan, C., Li, J. and Yang, Y. (2013) Spatial Distribution of the Extreme Hydrological Events in Xinjiang, North-West China. Natural Hazards, 67, 483-495. http://dx.doi.org/10.1007/s11069-013-0574-5
4. Wang, X., Xie, H. and Liang, T. (2007) Evaluation of MODIS Snow Cover and Cloud Mask and Its Application in Northern Xinjiang, China. Remote Sensing of Environment, 10, 1-17. http://dx.doi.org/10.1016/j.rse.2007.05.016
5. Al-Hanbali, A., Alsaaideh, B. and Kondoh, A. (2011) Using GIS-Based Weighted Linear Combination Analysis and Remote Sensing Techniques to Select Optimum Solid Waste Disposal Sites within Mafraq City, Jordan. Journal of Geographic Information System, 3, 1-12. http://dx.doi.org/10.4236/jgis.2011.34023
6. Sylvain, B., Anny, C., Nelly, M. M., William, L. and Frédéric, F. (2011) Satellite-Based High Latitude Snow Volume Trend, Variability and Contribution to Sea Level over 1989/2006. Global

and Planetary Change, 75, 99-107.http://dx.doi.org/10.1016/j.gloplacha.2010.10.011

7. Liang, T., Zhang, X., Xie, H., Wu, C., Feng, Q., Huang, X. and Chen, Q. (2008) Toward Improvely Daily Snow Cover Mapping with Advanced Combination of MODIS and AMSR-E Measurement. Remote Sensing Environment, 112, 3750-3761.http://dx.doi.org/10.1016/j.rse.2008.05.010

8. Dietza, A.J., Kuenzera, C. and Conrad, C. (2013) Snow-Cover Variability in Central Asia between 2000 and 2011 De- rived from Improved MODIS Daily Snow-Cover Products. International Journal of Remote Sensing, 34, 3879-3902.http://dx.doi.org/10.1080/01431161.2013.767480

9. Gao, Y., Xie, H., Lu, N., Yao, T. and Liang, T. (2010) Toward Advanced Daily Cloud-Free Snow Cover and Snow Water Equivalent Products Deom Terra-Aqua MODIS and Aqua AMSR-E Measurements. Journal of Hydrology, 385, 23-35.http://dx.doi.org/10.1016/j.jhydrol.2010.01.022

10. Liu, X., Liu, Y., Li, L. and Ren, Y. (2009) Disaster Monitoring and Early-Warning System for Snow Avalanche along Tianshan Highway. IGRARSS, II-634-637.

11. Liu, F., Mao, X., Zhang, Y., Chen, Q., Liu, P. and Zhao, Z. (2014) Risk Analysis of Snow Disaster in the Pastoral Areas of the Qinghai-Tibet Plateau. Journal of Geographical Science, 24, 411-426. http://dx.doi.org/10.1007/s11442-014-1097-z

Citations

CHAPTER 1

Thomas Wilson, Jim Cole, David Johnston, Shane Cronin, Carol Stewart, and Andre Dantas, Short- and Long-term Evacuation of People and Livestock during a Volcanic Crisis: Lessons from the 1991 Eruption of Volcán Hudson, Chile, doi:10.1186/2191-5040-1-2.

CHAPTER 2

Meihua Wu, Yaning Chen, Huaijun Wang, and Guili Sun, Characteristics of Meteorological Disasters and their Impacts on the Agricultural Ecosystems in the Northwest of China: a Case Study in Xinjiang, doi:10.1186/s40677-015-0015-8.

CHAPTER 3

Jacqueline McIntosh (2013). The Implications of Post Disaster Recovery for Affordable Housing, Approaches to Disaster Management - Examining the Implications of Hazards, Emergencies and Disasters, Prof. John Tiefenbacher (Ed.), ISBN: 978-953-51-1093-4, InTech, DOI: 10.5772/55273.

CHAPTER 4

Yoshii, H. , Saito, H. , Kikuchi, S. , Ueno, T. and Sato, K. (2014) Maternal Anxiety 16 Months after the Great East Japan Earthquake Disaster Area: First Report.Health,6, 870-878. doi:10.4236/health.2014.610110.

CHAPTER 5

Jun Tomio, Hajime Sato, Yuji Matsuda, Toshie Koga, Hiroko Mizumura, Household, and Community Disaster Preparedness in Japanese Provincial City: A Population-Based Household Survey, DOI:10.4236/aa.2014.42010.

CHAPTER 6

Wieslander, I. , Mårtensson, J. , Fridlund, B. and Svedberg, P. (2013) Factors Influencing Female Patients' Recovery after their first Myocardial Infarction as Experienced by Cardiac Rehabilitation Nurses. Open Journal of Nursing, 3,230-240. doi:10.4236/ojn.2013.32032.

CHAPTER 7

Zhenwang Li, Quanjun Jiao, Liangyun Liu, Huan Tang, and Tong Liu, Monitoring Geologic Hazards and Vegetation Recovery in the Wenchuan Earthquake Region Using Aerial Photography doi:10.3390/ijgi3010368.

CHAPTER 8

Erik R. Svendsen, Jennifer R. Runkle, Venkata Ramana Dhara, Shao Lin, Marina Naboka, Timothy A. Mousseau, and Charles Bennett, Epidemiologic Methods Lessons Learned from Environmental Public Health Disasters: Chernobyl, the World Trade Center, Bhopal, and Graniteville, South Carolina, doi:10.3390/ijerph9082894.

CHAPTER 9

Shimura, Y. and Yamamoto, K. (2014) Method of Searching for Earthquake Disaster Evacuation Routes Using Multi-Objective GA and GIS. Journal of Geographic Information System, 6, 492-525. doi: 10.4236/jgis.2014.65042.

CHAPTER 10

Abake, G., Al-Hanbali, A., Alsaaideh, B. and Tateishi, R. (2014) Potential Hazard Map for Snow Disaster Prevention Using GIS-Based Weighted Linear Combination Analysis and Remote Sensing Techniques: A Case Study in Northern Xinjiang, China. Advances in Remote Sensing, 3, 260-271. doi: 10.4236/ars.2014.34018.

Index

D

E

F

G

H

I

J

K

L

M

Index

D

E

F

G

H

I

J

K

L

M